Between Elementary and Advanced Mathematics I
- Interesting Topics not Covered in School/College Syllabus

z.c. Lin

Table of Contents

Preface v

1   The inequality $\sqrt[n]{x_1 \cdots x_n} \leq \frac{x_1+\cdots+x_n}{n}$ 1
2   How to sum $1^2 + \cdots + n^2$ 5
3   The infinite sum $1 + \dfrac{1}{2^2} + \dfrac{1}{3^2} + \cdots$ 7
4   The $n!$ approximation formula 11
5   Cubic equations 17
6.1   Basic properties of polynomials 25
6.2   Cubic equations without elementary solutions 31
7   Quartic equations 35
8    $\cos(2\pi/p)$ 39
9   Elementary properties of the prime numbers 49
10   The integer equation $x^2 + y^2 = z^2$ 55
11   The integer equation $x^3 + y^3 = z^3$ 63
12   The Fibonacci numbers 75
13   Continued fractions 77

Contents of Part II 83

Index

Preface

In mathematics, there are many topics that cannot be found in school and university standardized textbooks. Those topics are ignored mostly because they are not considered to be foundational mathematics, and are not usually needed in science and technology.

It's a pity that some important and really nice topics are ignored as well. And as a result, many students are unaware of these nice results and the equally interesting techniques behind them.

The aim of this book is to introduce to the reader a few of such interesting mathematical topics that are normally ignored by textbooks.

I have managed to make the sections more or less self-contained, so that the reader can jump directly to a topic without being bothered by the rest of the book.

Some of the topics require basics of calculus. §6 also uses basic concepts of linear algebra.

Included at the end is a table of contents of a sequel to this book, which discusses 2-dimensional surfaces.

1 The inequality $\sqrt[n]{x_1 \cdots x_n} \leq \frac{x_1 + \cdots + x_n}{n}$

This inequality states that: $\sqrt[n]{x_1 \cdots x_n} \leq \frac{x_1 + \cdots + x_n}{n}$, where $x_1, \ldots, x_n$ are positive numbers. And the left and right sides are equal only when $x_1 = \cdots = x_n$.

Proof 1

A simple way of proving this inequality consists of the following three steps. Full details will be left to the reader as an exercise.

(1) Verify that the inequality is true when $n = 2$, i.e., $\sqrt{ab} \leq \frac{a+b}{2}$.
(2) Show that if the inequality is true for n, then it must be true for $2n$.
(3) Show that if the inequality is true for $(n+1)$, then it must be true for n:
$$(x_1 \cdots x_n \cdot y)^{\frac{1}{n+1}} \leq \frac{x_1 + \cdots + x_n + y}{n+1}$$
Let $G = \sqrt[n]{x_1 \cdots x_n}$, $A = \frac{x_1 + \cdots + x_n}{n}$. Then the above becomes
$$(G^n y)^{\frac{1}{n+1}} \leq \frac{nA + y}{n+1}$$
Let $y = G$, or $y = A$, then the above gives $G \leq A$. This proves (3).

Exercise: Explain why the above three steps together imply the inequality for an arbitrary positive integer n. By going through the steps, explain why "$\leq$" becomes "$=$" only when $x_1 = \cdots = x_n$.

Proof 2

(1) $(1+x)^n \leq 1 + nx$, $x \geq 0$, $0 \leq n \leq 1$.

To see why, compare the graphs of functions $f(x) = (1+x)^n$ and $g(x) = 1 + nx$. When $x = 0$, they have the same value($= 1$). But their slopes are $f'(x) = n(1+x)^{n-1} \leq n$, $g'(x) = n$. So the graph of $f(x)$ will always be below that of $g(x)$. So $f(x) \leq g(x)$, $x \geq 0$.

(2) $x_1^{\lambda_1} \cdots x_n^{\lambda_n} \leq \lambda_1 x_1 + \cdots + \lambda_n x_n$, where $x_1, \cdots, x_n$ are positive numbers, $\lambda_1, \cdots, \lambda_n$ are positive numbers, $\lambda_1 + \cdots + \lambda_n = 1$.

To prove it via induction, we first reduce the case $n = 2$ to (1). Now we need to verify that if (2) is true for n, then it must also be true for $n+1$. So assume that $x_1, \cdots, x_n, y; \lambda_1, \cdots, \lambda_n, \mu$ are all positive numbers, $\lambda = \lambda_1 + \cdots + \lambda_n$, $\lambda + \mu = 1$.

$$x_1^{\lambda_1} \cdots x_n^{\lambda_n} y^\mu = (x_1^{\lambda_1/\lambda} \cdots x_n^{\lambda_n/\lambda})^\lambda y^\mu \leq (\tfrac{\lambda_1}{\lambda} x_1 + \cdots + \tfrac{\lambda_n}{\lambda} x_n)^\lambda y^\mu$$
$$\leq \lambda(\tfrac{\lambda_1}{\lambda} x_1 + \cdots + \tfrac{\lambda_n}{\lambda} x_n) + \mu y$$
$$= \lambda_1 x_1 + \cdots + \lambda_n x_n + \mu y$$

(3) The inequality $\sqrt[n]{x_1 \cdots x_n} \leq \frac{x_1 + \cdots + x_n}{n}$, where $x_1, \ldots, x_n$ are positive numbers, is obviously a special case of (2). #[*]

Proof 3

This is just a slightly more general version of the second proof.

(1) A function is said to be (upward) convex if its graph looks like $\smile$.

Exercise: Explain why a function $y = f(x)$ is convex if and only if $f''(x) \geq 0$ everywhere, by using the fact that, $f'(x)$ represents the slope of the curve $y = f(x)$ at point (x, y).

(2) Let $y = f(x)$ be a convex function. Draw a graph to explain why $f(\lambda x + \mu y) \leq \lambda f(x) + \mu f(y)$, where $\lambda, \mu > 0$, $\lambda + \mu = 1$.

---

[*]In this book, the symbol "#" is used to inform the reader that the proof of a statement has been completed.

(3) Let $y = f(x)$ be a convex function. Deduce that
$f(\lambda_1 x_1 + \cdots + \lambda_n x_n) \leq \lambda_1 f(x_1) + \cdots + \lambda_n f(x_n)$,
where $\lambda_1, \cdots, \lambda_n$ are positive, $\lambda_1 + \cdots + \lambda_n = 1$.

(4) Explain why and how the exponential function e^x can be to deduce the inequality $\sqrt[n]{x_1 \cdots x_n} \leq \dfrac{x_1 + \cdots + x_n}{n}$.

An Application

(1) From the inequality, deduce that
$$n! < \left(\frac{n+1}{2}\right)^n$$

(2) Note that in inequality $\sqrt[n]{x_1 \cdots x_n} \leq \dfrac{x_1 + \cdots + x_n}{n}$, the difference between the two sides should be smaller, when the differences among $\{x_1, \cdots, x_n\}$ are not too great. Use a set of numbers $\{x_1, \cdots, x_n\}$, in which differences are smaller than in the set $\{1, 2, \cdots, n\}$, to show that
$$n! < \left(\frac{n}{2}\right)^n, n \geq 6$$

2 How to sum $1^2 + \cdots + n^2$

Here is how we calculate $1^2 + \cdots + n^2$:

Let's look at $(n+1)^3 - n^3 = 3n^2 + 3n + 1$
Let n be $1, 2, \cdots, n$
$$2^3 - 1^3 = 3 \cdot 1^2 + 3 \cdot 1 + 1$$
$$3^3 - 2^3 = 3 \cdot 2^2 + 3 \cdot 2 + 1$$
$$4^3 - 3^3 = 3 \cdot 3^2 + 3 \cdot 3 + 1$$
$$5^3 - 4^3 = 3 \cdot 4^2 + 3 \cdot 4 + 1$$
$$\cdots$$
$$(n+1)^3 - n^3 = 3n^2 + 3n + 1$$
Adding all the above identities together gives
$(n+1)^3 - 1^3 = 3(1^2 + \cdots + n^2) + 3(1 + \cdots + n) + n$
Therefore
$(n+1)^3 - 1 = 3(1^2 + \cdots + n^2) + 3 \cdot \frac{n(n+1)}{2} + n$
From this we obtain the following formula
$$1^2 + \cdots + n^2 = \frac{n(n+1)(2n+1)}{6}$$

Exercise:
Use $(n+1)^4 - n^4 = 4n^3 + 6n^2 + 4n + 1$
to show that
$$1^3 + \cdots + n^3 = \left(\frac{n(n+1)}{2}\right)^2$$

3 The infinite sum $1 + \dfrac{1}{2^2} + \dfrac{1}{3^2} + \cdots$

We are interested in finding out what the infinite sum $1 + \frac{1}{2^2} + \frac{1}{3^2} + \cdots$ should be equal to. It turns out we can use calculus to identify it as $\pi^2/6$. Our "proof" here may not be 100% logically sound. But our point here is to relate the infinite sum to something, which turns out to be $\pi^2/6$. We can then (say) use a calculator to convince ourselves that it's the correct answer.

Recall that
$\sin(x) = x - x^3/3! + x^5/5! - \cdots$

Let's think of $\sin(x)$ as a "polynomial of degree=∞". The roots of this "polynomial" are $x = 0, \pm\pi, \pm 2\pi, \pm 3\pi, \cdots$. Therefore one must be able to express $\sin(x)$ as

$\sin(x) = A \cdot x \cdot (x+\pi)(x-\pi) \cdot (x+2\pi)(x-2\pi) \cdot (x+3\pi)(x-3\pi) \cdots$
$= A \cdot x \cdot [x^2 - \pi^2] \cdot [x^2 - (2\pi)^2] \cdot [x^2 - (3\pi)^2] \cdots,$

where A is a constant.
Rearrange the above as

$\sin(x) = B \cdot x \cdot \left[1 - \left(\dfrac{x}{\pi}\right)^2\right] \cdot \left[1 - \left(\dfrac{x}{2\pi}\right)^2\right] \cdot \left[1 - \left(\dfrac{x}{3\pi}\right)^2\right] \cdots,$

where B is a constant. But recall that, $\sin(x)/x \to 1$ as $x \to 0$. Therefore, the constant B must be 1.

We have thus established another expression of $\sin(x)$ as

$\sin(x) = x \cdot \left[1 - \left(\dfrac{x}{\pi}\right)^2\right] \cdot \left[1 - \left(\dfrac{x}{2\pi}\right)^2\right] \cdot \left[1 - \left(\dfrac{x}{3\pi}\right)^2\right] \cdots$

Let's now compare these two expressions of $\sin(x)$.
Firstly let's expand the product, a product of infinite many terms,

$$x \cdot \left[1 - \left(\frac{x}{\pi}\right)^2\right] \cdot \left[1 - \left(\frac{x}{2\pi}\right)^2\right] \cdot \left[1 - \left(\frac{x}{3\pi}\right)^2\right] \cdots$$

In the expansion, the first term is obviously x.
The next term is an x^3-term, which is

$$x \cdot \left[-\left(\frac{x}{\pi}\right)^2 - \left(\frac{x}{2\pi}\right)^2 - \left(\frac{x}{3\pi}\right)^2 - \cdots\right] = -\left(1 + \frac{1}{2^2} + \frac{1}{3^2} + \cdots\right) \cdot \frac{1}{\pi^2} \cdot x^3$$

(The term after is an x^5-term, etc)
Therefore $\sin(x) = x - \left(1 + \frac{1}{2^2} + \frac{1}{3^2} + \cdots\right) \cdot \frac{1}{\pi^2} \cdot x^3 + \cdots$
Compare this with $\sin(x) = x - x^3/3! + x^5/5! - \cdots$
So $\left(1 + \frac{1}{2^2} + \frac{1}{3^2} + \cdots\right) \cdot \frac{1}{\pi^2} = 1/3!$
So $1 + \frac{1}{2^2} + \frac{1}{3^2} + \cdots = \pi^2/6$

<u>Exercise</u>: $1 + \frac{1}{2^4} + \frac{1}{3^4} + \cdots = \frac{\pi^4}{90}$

This comes from comparing the coefficients of the x^5-terms of the two expressions of $\sin(x)$.

For $\sin(x) = x - x^3/3! + x^5/5! - \cdots$
The term is $x^5/5!$
For $\sin(x) = x \cdot \left[1 - \left(\frac{x}{\pi}\right)^2\right] \cdot \left[1 - \left(\frac{x}{2\pi}\right)^2\right] \cdot \left[1 - \left(\frac{x}{3\pi}\right)^2\right] \cdots$
The term is

$$x \cdot \sum_{n<m} (-1)\left(\frac{x}{n\pi}\right)^2 (-1)\left(\frac{x}{m\pi}\right)^2 = \left(\sum_{n<m} \frac{1}{n^2}\frac{1}{m^2}\right) \frac{1}{\pi^4} \cdot x^5$$

So

$$\left(\sum_{n<m} \frac{1}{n^2}\frac{1}{m^2}\right) \frac{1}{\pi^4} = 1/5! \implies \sum_{n<m} \frac{1}{n^2}\frac{1}{m^2} = \pi^4/120$$

Now consider

$$\left(\sum_{n=1}^{\infty}\frac{1}{n^2}\right)\left(\sum_{m=1}^{\infty}\frac{1}{m^2}\right) = \sum_{n,m}\frac{1}{n^2}\frac{1}{m^2} = \sum_{n=m}+2\sum_{n<m}$$
$$= \left(1+\frac{1}{2^4}+\frac{1}{3^4}+\cdots\right)+2\left(\pi^4/120\right)$$

This implies

$$\left(\pi^2/6\right)^2 = \left(1+\frac{1}{2^4}+\frac{1}{3^4}+\cdots\right)+\pi^4/60$$

Therefore

$$1+\frac{1}{2^4}+\frac{1}{3^4}+\cdots = (1/36-1/60)\pi^4 = \pi^4/90$$

Exercise: Prove the following

$$\sum_{n=1}^{\infty}\frac{1}{n^2}\left(\frac{1}{2}\right)^n = \frac{1}{2}\left[\frac{\pi^2}{6}-\ln(2)^2\right]$$

The proof

Let $y = \sum_{n=1}^{\infty}\frac{1}{n^2}x^n$

$(xy')' = 1+x+x^2+\cdots = \dfrac{1}{1-x}$

$xy' = -\ln(1-x)$

$$y = \int_0^x -\frac{\ln(1-x)}{x}dx$$
$$= \int_0^x -\ln(1-x)d(\ln x) = -\ln(x)\ln(1-x)|_0^x - \int_0^x \frac{\ln(x)}{1-x}dx$$

$\lim_{x\to 0^+}\ln(x)\ln(1-x) = \cdots = 0$

$$\int_0^x \frac{\ln(x)}{1-x}dx = \int_0^x \frac{\ln[1-(1-x)]}{1-x}(-1)d(1-x) = \int_1^{1-x} -\frac{\ln(1-x)}{x}dx$$
$$= y(1-x) - y(1).$$

Thus, $y(x) = -\ln(x)\ln(1-x) + y(1) - y(1-x)$

$y(x) + y(1-x) = \pi^2/6 - \ln(x)\ln(1-x)$
Let $x = 1/2$,
$2y(1/2) = \pi^2/6 - (\ln 2)^2$
$y(1/2) = \frac{1}{2}[\frac{\pi^2}{6} - (\ln 2)^2]$

4 The $n!$ approximation formula

We want to prove the following formula

$$n! \sim \sqrt{2\pi} \cdot \frac{n^{n+1/2}}{e^n}$$

i.e., Left/Right approaches 1 when $n \to \infty$.

The idea is that $n! = e^{\ln(n!)} = e^{\ln(1)+\cdots+\ln(n)}$, and as we will see, $\ln(1) + \ln(2) + \cdots + \ln(n)$ can be estimated.

In general, given a function $f(x)$, there is no formula for the sum $f(1) + \cdots + f(n)$. Yet we will see that the sum can usually be estimated. We have seen in §2 that, in order to determine $1^2 + \cdots n^2$, we make use of the function n^3, i.e., via identity $(n+1)^3 - n^3 = 3n^2 + 3n + 1$, by letting $n = 1, 2, \cdots, n$, and then adding everything together. And in order to determine $1^3 + \cdots + n^3$, we make use of function n^4, etc. But for a general function $f(x)$, to determine, or at least to estimate $f(1) + \cdots + f(n)$, what associated function must we use? The answer is, use $F(x) = \int f(x) dx$. Let's try an example first.

4.1 Example: $1 + \dfrac{1}{2} + \cdots + \dfrac{1}{n} = \ln(n) +$ Convergent.

The word "Convergent" means a sequence that is convergent as $n \to \infty$.

<u>Proof:</u> $\displaystyle\int \frac{1}{x} dx = \ln(x) +$ Constant, so we try function $F(x) = \ln(x)$.
$\ln(n+1) - \ln(n) = \ln(1 + \frac{1}{n})$
From $\dfrac{1}{1+x} = 1 - x + x^2 - x^3 + \cdots$

So $\ln(1+x) = x - \dfrac{x^2}{2} + \dfrac{x^3}{3} - \dfrac{x^4}{4} + \cdots$

Let $x = 1/n$

$\ln(n+1) - \ln(n) = \dfrac{1}{n} - \dfrac{1}{2}\dfrac{1}{n^2} + \dfrac{1}{3}\dfrac{1}{n^3} - \dfrac{1}{4}\dfrac{1}{n^4} + \cdots$

Set $n = 1, 2, \cdots, n$, then add together all the identities obtained

$$\ln(n+1) - \ln(1) = \left(1 + \dfrac{1}{2} + \cdots + \dfrac{1}{n}\right)$$
$$-\dfrac{1}{2}\left(1 + \dfrac{1}{2^2} + \cdots + \dfrac{1}{n^2}\right)$$
$$+\dfrac{1}{3}\left(1 + \dfrac{1}{2^3} + \cdots + \dfrac{1}{n^3}\right)$$
$$-\dfrac{1}{4}\left(1 + \dfrac{1}{2^4} + \cdots + \dfrac{1}{n^4}\right)$$
$$+\cdots$$

Take a look at

$$-\dfrac{1}{2}\left(1 + \dfrac{1}{2^2} + \cdots + \dfrac{1}{n^2}\right) + \dfrac{1}{3}\left(1 + \dfrac{1}{2^3} + \cdots + \dfrac{1}{n^3}\right) - \dfrac{1}{4}\left(1 + \dfrac{1}{2^4} + \cdots + \dfrac{1}{n^4}\right) + \cdots$$

It is an infinite sum of an alternating sequence. $1 + \frac{1}{2^2} + \cdots + \frac{1}{n^2}$, $1 + \frac{1}{2^3} + \cdots + \frac{1}{n^3}$, $1 + \frac{1}{2^4} + \cdots + \frac{1}{n^4}$, etc, have a common upper bound. Therefore the limit of the alternating sequence is zero. Thus this infinite sum is convergent.

In other words

$$\ln(n+1) - \ln(1) = 1 + \dfrac{1}{2} + \cdots + \dfrac{1}{n} + \text{Convergent}.$$

Note that $\ln(n+1) = \ln(n) + \ln(1 + 1/n)$, $\ln(1 + 1/n)$ is also convergent.

This proves our claim that $1 + \dfrac{1}{2} + \cdots \dfrac{1}{n} = \ln(n) + \text{Convergent}$.

Reminder: The standard procedure that demonstrates that
$$1 + 1/2 + 1/3 + \cdots = +\infty$$
is as follows:

$1 + 1/2 + 1/3 + \cdots$
$= 1 + 1/2 + (1/3 + 1/4) + (1/5 + 1/6 + 1/7 + 1/8) + (1/9 + \cdots + 1/16) + \cdots$
$\geq 1 + 1/2 + (1/4 + 1/4) + (1/8 + 1/8 + 1/8 + 1/8) + (1/16 + \cdots + 1/16) + \cdots$
$= 1 + 1/2 + 2 \cdot 1/4 + 4 \cdot 1/8 + 8 \cdot 1/16 + \cdots$
$= 1 + 1/2 + 1/2 + 1/2 + 1/2 + \cdots$
$= +\infty$

But unlike the precise estimate, this simple procedure can't tell us exactly how the sum grows toward $+\infty$.

4.2 Let's now estimate $\ln(1) + \cdots + \ln(n)$

Recall that $\int \ln(x) dx = x \ln(x) - x + \text{Constant}$
Let $F(x) = x \ln(x) - x$
$F(n+1) - F(n) = [(n+1)\ln(n+1) - (n+1)] - [n \ln(n) - n]$
$$= n \ln(1 + \frac{1}{n}) + \ln(n+1) - 1$$
$$= n \left[\frac{1}{n} - \frac{1}{2}\frac{1}{n^2} + \frac{1}{3}\frac{1}{n^3} - \frac{1}{4}\frac{1}{n^4} + \cdots \right] + \ln(n+1) - 1$$
$$= \left(-\frac{1}{2}\frac{1}{n} + \frac{1}{3}\frac{1}{n^2} - \frac{1}{4}\frac{1}{n^3} + \cdots \right) + \ln(n+1)$$

Set $n = 1, 2, \cdots, n$. Then sum up all the identities obtained

$$F(n+1) - F(1) = -\frac{1}{2}\left(1 + \cdots + \frac{1}{n}\right)$$
$$+ \frac{1}{3}\left(1 + \cdots + \frac{1}{n^2}\right)$$
$$- \frac{1}{4}\left(1 + \cdots + \frac{1}{n^3}\right)$$
$$+ \cdots$$
$$+ [\ln(2) + \cdots + \ln(n+1)]$$

The right hand side of this equation:
According to Example 4.1, $1 + \cdots + \frac{1}{n} = \ln(n) + \text{Convergent}$.
As also explained there, the following infinite sum is convergent
$$-\frac{1}{2}\left(1 + \cdots + \frac{1}{n^2}\right) + \frac{1}{3}\left(1 + \cdots + \frac{1}{n^3}\right) - \cdots$$
In conclusion, the right hand side is
$-\frac{1}{2}\ln(n) + [\ln(2) + \cdots + \ln(n+1)] + \text{Convergent}$.

The left hand side of the equation:
$F(n+1) - F(1) = [(n+1)\ln(n+1) - (n+1)] - [-1] = (n+1)\ln(n+1) - n$
$(n+1)\ln(n+1) = n \cdot \ln(n+1) + \ln(n+1)$
$n \cdot \ln(n+1) = n \cdot \ln(n) + n \cdot \ln(1 + 1/n)$
$$n \cdot \ln(1 + 1/n) = \frac{\ln(1 + 1/n)}{1/n} \to \lim_{x \to 0} \frac{\ln(1+x)}{x} = 1,$$
so is convergent.
In conclusion, the left hand side is
$n \cdot \ln(n) + \ln(n+1) - n + \text{Convergent}$

We thus end up with the following asymptotic formula
$$\ln(1) + \cdots + \ln(n) = (n + \frac{1}{2})\ln(n) - n + \text{Convergent}$$

4.3 The proof of the approximation formula

$$n! \sim \sqrt{2\pi} \cdot \frac{n^{n+1/2}}{e^n}$$

The asymptotic formula established in 4.2 implies $n! = C_n \cdot \frac{n^{n+1/2}}{e^n}$, where C_n is convergent as $n \to \infty$. We now show that C_n is convergent to $\sqrt{2\pi}$.

According to §3,
$$\sin(x) = x \cdot \left[1 - \left(\frac{x}{\pi}\right)^2\right] \cdot \left[1 - \left(\frac{x}{2\pi}\right)^2\right] \cdot \left[1 - \left(\frac{x}{3\pi}\right)^2\right] \cdots$$

Let $x = \pi/2$,

$$1 = \frac{\pi}{2} \cdot [1 - (\frac{1}{2})^2] \cdot [1 - (\frac{1}{4})^2] \cdot [1 - (\frac{1}{6})^2] \cdots$$

$$[1 - (\frac{1}{2})^2] \cdot [1 - (\frac{1}{4})^2] \cdot [1 - (\frac{1}{6})^2] \cdots [1 - (\frac{1}{2n})^2]$$
$$= \frac{1 \times 3}{2^2} \cdot \frac{3 \times 5}{4^2} \cdot \frac{5 \times 7}{6^2} \cdots \frac{(2n-1) \times (2n+1)}{(2n)^2}$$
$$= \frac{[1 \times 3 \times 5 \times \cdots \times (2n-1)]^2 (2n+1)}{(2 \times 4 \times 6 \times \cdots \times 2n)^2}$$
$$= \frac{[(2n)!]^2 (2n+1)}{(2 \times 4 \times 6 \times \cdots \times 2n)^4}$$
$$= \frac{[(2n)!]^2 (2n+1)}{(2^n \cdot n!)^4}$$

Now $n! = C_n \cdot \dfrac{n^{n+1/2}}{e^n}$. Assume $C_n \to C$ as $n \to \infty$.

$$\frac{[(2n)!]^2(2n+1)}{(2^n \cdot n!)^4} \longrightarrow \frac{[C \cdot \frac{(2n)^{2n+1/2}}{e^{2n}}]^2 (2n+1)}{[2^n \cdot C \cdot \frac{n^{n+1/2}}{e^n}]^4} = \frac{1}{C^2} \cdot 2 \cdot \frac{1}{n} \cdot (2n+1) \longrightarrow \frac{4}{C^2}$$

$$1 = \frac{\pi}{2} \cdot \frac{4}{C^2} \implies C = \sqrt{2\pi}$$

5 Cubic equations

5.1 The standard procedure for solving a cubic equation

Given a general cubic equation
$x^3 + ax^2 + bx + c = 0$
The first step in the procedure is to eliminate the x^2-term,
by letting $y = x + a/3$.

> If the three roots of the equation are $x = \alpha, \beta, \gamma$, then
> $\alpha + \beta + \gamma = -a$.
> So $(\alpha + a/3) + (\beta + a/3) + (\gamma + a/3) = 0$.
> By letting $y = x + a/3$, the sum of the three roots of the y-equation is zero.
> So the new equation has no y^2-term.

The equation now looks like
$x^3 + px + q = 0$

> If the x-term can be further eliminated, then $x = \sqrt[3]{\square}$. The procedure would be comparable to the way we solve quadratic equations.
>
> In general, the x-term cannot be eliminated. So we cannot expect the solution to look like $x = \sqrt[3]{\square}$.
>
> Instead, as we will see now, it turns out that the solution looks like
> $$x = \sqrt[3]{\boxed{1}} + \sqrt[3]{\boxed{2}}$$

To continue:
Let $x = u + v$
$(u+v)^3 + p(u+v) + q = 0$
$u^3 + 3uv(u+v) + v^3 + p(u+v) + q = 0$
$(u^3 + v^3 + q) + (3uv + p)(u+v) = 0$
Let
$$\begin{cases} u^3 + v^3 + q = 0 \\ 3uv + p = 0 \end{cases}$$

This leads to a quadratic equation, whose roots are u^3 and v^3. Solve it for u^3 and v^3. Then determine u and v.

Example: $(x+1)(x+2)(x+3) = 10$

This equation has one real solution (and two complex ones), if you take a look at its graph. Let's find the real one.

$x^3 + 6x^2 + 11x - 4 = 0$
Let $y = x + 2$
$x = y - 2$
$y^3 - y - 10 = 0$

Let $y = u + v$
$u^3 + 3uv(u+v) + v^3 - (u+v) - 10 = 0$
$(u^3 + v^3 - 10) + (3uv - 1)(u+v) = 0$
Let
$$\begin{cases} u^3 + v^3 = 10 \\ uv = 1/3 \end{cases}$$
So
$$\begin{cases} u^3 + v^3 = 10 \\ u^3 v^3 = 1/27 \end{cases}$$

The quadratic equation whose roots are u^3 and v^3 is
$z^2 - 10z + 1/27 = 0$

$$z = 5 \pm \sqrt{\frac{674}{27}}$$

$$u^3 = 5 + \sqrt{\frac{674}{27}}, \quad v^3 = 5 - \sqrt{\frac{674}{27}}$$

$$x = \sqrt[3]{5 + \sqrt{\frac{674}{27}}} + \sqrt[3]{5 - \sqrt{\frac{674}{27}}} - 2 \approx 0.3089$$

5.2 $x^3 - 3x + 1 = 0$

In a sense, cubic equations can be classified in terms of a very small number of types. This equation represents the type that is the most important/interesting.

Solution of 5.2 using trigonometry

$\sin 3\theta = 3\sin\theta - 4\sin^3\theta$
Let $\theta = 10°$
$1/2 = 3\sin\theta - 4\sin^3\theta$
$8\sin^3\theta - 6\sin\theta + 1 = 0$
$(2\sin\theta)^3 - 3(2\sin\theta) + 1 = 0$
Let $x = 2\sin 10°$
$x^3 - 3x + 1 = 0$
Let $\theta = 50° \longrightarrow x = 2\sin 50°$
Let $\theta = 70° \longrightarrow x = -2\sin 70°$

We conclude that the roots of $x^3 - 3x + 1 = 0$ are $2\sin 10°$, $2\sin 50°$, $-2\sin 70°$.

Corollary

$$\sin 10° + \sin 50° = \sin 70°$$
$$\sin 10° \cdot \sin 50° \cdot \sin 70° = 1/8$$
$$\frac{1}{\sin 10°} + \frac{1}{\sin 50°} - \frac{1}{\sin 70°} = 6$$

Direct proof that $\sin 10° + \sin 50° = \sin 70°$

$\sin 70° = \sin(60° + 10°) = \frac{\sqrt{3}}{2}\cos(10°) + \frac{1}{2}\sin 10°$
$\sin 50° = \sin(60° - 10°) = \frac{\sqrt{3}}{2}\cos(10°) - \frac{1}{2}\sin 10°$
$\sin 70° - \sin 50° = \sin 10°$

Solution of 5.2 using our cubic equation procedure

$x^3 - 3x + 1 = 0$
Let $x = u + v$
$u^3 + 3uv(u+v) + v^3 - 3(u+v) + 1 = 0$
$(u^3 + v^3 + 1) + (3uv - 3)(u+v) = 0$
Let
$$\begin{cases} u^3 + v^3 = -1 \\ uv = 1 \end{cases}$$

The quadratic equation whose roots are u^3 and v^3 is
$z^2 + z + 1 = 0$
$$z = \frac{-1 \pm \sqrt{3}i}{2}$$

$$\frac{-1 + \sqrt{3}i}{2} = e^{i120°}$$

$u^3 = e^{i120°}$

$u = e^{i(120° + 360°n)/3} = e^{i(40° + 120°n)}, n = 0, 1, 2$
$\quad = e^{i40°}, e^{i160°}, e^{i280°}$

Note that $uv = 1$, so $v = e^{-i40°}, e^{-i160°}, e^{-i280°}$

$x = u + v = 2\cos 40° = 2\sin 50°$
$\qquad 2\cos 160° = -2\cos 20° = -2\sin 70°$
$\qquad 2\cos 280° = 2\cos 80° = 2\sin 10°$

Exercise: Show that $\sin 10°$ is not a rational number.

5.3 $x^3 + 3x + 1 = 0$

It turns out that this equation is completely different from its "similar" version $x^3 - 3x + 1 = 0$. The function $f(x) = x^3 + 3x + 1$ is increasing because $f'(x) = 3(x^2 + 1) > 0$, and goes from $-\infty$ to $+\infty$. So the equation $x^3 + 3x + 1 = 0$ has exactly one real root.

Let's find this real root
Let $x = u + v$
$u^3 + 3uv(u + v) + v^3 + 3(u + v) + 1 = 0$
$(u^3 + v^3 + 1) + (3uv + 3)(u + v) = 0$
Let
$$\begin{cases} u^3 + v^3 = -1 \\ uv = -1 \end{cases}$$

The quadratic equation whose roots are u^3 and v^3 is
$z^2 + z - 1 = 0$
$$z = \frac{-1 \pm \sqrt{5}}{2}$$

$$u,\ v = \sqrt[3]{\frac{-1 + \sqrt{5}}{2}},\ \sqrt[3]{\frac{-1 - \sqrt{5}}{2}}$$

$$x = -\left(\sqrt[3]{\frac{\sqrt{5} + 1}{2}} - \sqrt[3]{\frac{\sqrt{5} - 1}{2}}\right) \approx -0.3222$$

5.4 The discriminant of a cubic equation

In general, suppose $P(x)$ is a polynomial of degree n, whose roots are $\alpha_1, \cdots, \alpha_n$. Then its discriminant is defined to be

$$\Delta = \left[\prod_{i<j}(\alpha_i - \alpha_j)\right]^2$$

Example:

Let's try the quadratic equation $ax^2 + bx + c = 0$
Let its roots be α, β

$$\begin{aligned}\Delta &= (\alpha - \beta)^2 = (\alpha + \beta)^2 - 4\alpha\beta \\ &= (-b/a)^2 - 4c/a \\ &= \frac{b^2 - 4ac}{a^2}\end{aligned}$$

Example:

For $x^3 + px + q = 0$,
$\Delta = -(4p^3 + 27q^2)$

Proof:

Let the roots be α, β, γ
$\Delta = [(\alpha - \beta)(\alpha - \gamma)(\beta - \gamma)]^2$
$(\alpha - \beta)^2 = (\alpha + \beta)^2 - 4\alpha\beta = (-\gamma)^2 - 4(-q/\gamma) = \gamma^2 + 4q/\gamma$

$$= \frac{\gamma^3 + 4q}{\gamma} = \frac{-p\gamma - q + 4q}{\gamma} = \frac{3q - p\gamma}{\gamma}$$

$$\begin{aligned}\Delta &= \frac{(3q - p\alpha)(3q - p\beta)(3q - p\gamma)}{\alpha\beta\gamma} \\ &= \frac{p^3}{-q}\left(\frac{3q}{p} - \alpha\right)\left(\frac{3q}{p} - \beta\right)\left(\frac{3q}{p} - \gamma\right) \\ &= -\frac{p^3}{q}\left[\left(\frac{3q}{p}\right)^3 + p\left(\frac{3q}{p}\right) + q\right] = -(4p^3 + 27q^2)\end{aligned}$$

Special cases:

$x^3 - 3x + 1 = 0:\quad \Delta = 3^4$
$x^3 + 3x + 1 = 0:\quad \Delta = -3^3 \times 5$

Example:

$x^3 + x^2 + x + 1 = 0$

> Determine Δ by using formula:
> Let $y = x + 1/3$
> $(y - 1/3)^3 + (y - 1/3)^2 + (y - 1/3) + 1 = y^3 + (2/3)y + 20/27$
> $\Delta = -[4p^3 + 27q^2] = -[4(2/3)^3 + 27(20/27)^2] = -2^4$
>
> Determine Δ via definition:
> $x^3 + x^2 + x + 1 = (x + 1)(x^2 + 1)$
> Its roots are $-1, \pm i$
> $\Delta = [(-1 - i)(-1 - -i)(i - -i)]^2 = [(2)(2i)]^2 = -2^4$

Exercises:

(1) For a cubic equation with real coefficients. Explain why
$\Delta = 0 \implies$ repeat roots
$\Delta > 0 \implies$ 3 distinct real roots
$\Delta < 0 \implies$ 1 real, 2 complex roots

(2) Use Calculus to show that equation $x^3 + px + q = 0$ has repeat roots if and only if $4p^3 + 27q^2 = 0$.

(3) Consider cubic equation
$$x^3 + ax^2 + bx + c = 0, \quad a, b, c \in \mathbb{Q}$$
Show that a necessary condition for the equation to have repeat roots is that
$$a^2 - 3b = (\text{rational})^2$$

6.1 Basic properties of polynomials

A real number is said to be "elementary" if it can be constructed from rational numbers through $+, -, \times, \div$, plus real radical operations. In the next section, we want to show that the equation $x^3 - 3x + 1 = 0$ has no "elementary solution", meaning that none of its three roots, all real, can be elementarily constructed. We really need the concept "field", without which "elementary construction" can't be precisely defined.

This section requires basic linear algebra concepts.

In mathematics, $\mathbb{Z}$ is the set of all integers, $\mathbb{Q}$ is the set of all rational numbers, $\mathbb{R}$ is the set of all real numbers, and $\mathbb{C}$ is the set of all complex numbers. Notice that in $\mathbb{Q}$, we have the four basic operations, $+, -, \times, \div$. The same is true in $\mathbb{R}$ and $\mathbb{C}$. But in $\mathbb{Z}$, we only have three operations, $+, -, \times$. The operation $\div$ doesn't work in $\mathbb{Z}$, since the division of an integer by another may no longer be integral. For example, 2 and 3 are integer, but 2/3 no longer is. $\mathbb{Q}, \mathbb{R}, \mathbb{C}$ are called "fields". $\mathbb{Z}$ is not a "field".

Definition: A set of numbers, within which the four basic operations $+, -, \times, \div$ still work, is called a field.

Let F be a field. A polynomial over F (a F-polynomial), is a certain
$$ax^n + bx^{n-1} + \cdots + c$$
where the coefficients $a, b, \cdots, c$ are all in F and the leading coefficient $a \neq 0$. n is called the degree of the polynomial. The set of all F-polynomials is denoted as $F[x]$.

For example, rational polynomials are polynomials whose coefficients are all rational numbers. $\mathbb{Q}[x]$ is the set of all rational polynomials.

Proposition: Suppose $P(x)$ and $Q(x)$ are polynomials, over a certain field, with greatest common divisor $D(x)$ (also a polynomial). Then there exist polynomials $A(x)$ and $B(x)$ such that $AP + BQ = D$.

This proposition is a result of the so-called Euclid algorithm in arithmetics, and can be proven via mathematical induction as follows: Assume, e.g., $\deg(P) \geq \deg(Q)$. By attempting to divide P by Q, we end up with a certain expression $P/Q = \square + R/Q$, where both $\square(x)$ and $R(x)$ are polynomials, with the degree of R lower than that of Q. Anyway, $P - \square \cdot Q$ is a polynomial of a degree lower than that of P. The greatest common divisor of $P - \square \cdot Q$ and Q is still D. Suppose the proposition were true for the pair $P - \square \cdot Q$ and Q, then,
$\text{Polynomial}_1 \cdot (P - \square \cdot Q) + \text{Polynomial}_2 \cdot (Q) = D$
$\text{Polynomial}_1 \cdot (P) + (\text{Polynomial}_2 - \text{Polynomial}_1 \cdot \square) \cdot (Q) = D$
The above means the proposition would also be true for the pair P and Q. Repeat the procedure again and again until one of the two polynomials becomes zero. Say $Q(x) \equiv 0$. Then $D(x) = P(x)$. In such a case the proposition is trivial. #

Over a field F, a F-polynomial is said to be irreducible if it can no longer be split into the product of some other (non trivial) F-polynomials.

Proposition: Different irreducible polynomials can't have common roots.

Suppose $P(x)$ and $Q(x)$ are different irreducible F-polynomials. Obviously, their greatest common divisor is 1. Therefore, there exist polynomials $A(x)$ and $B(x)$ such that $A(x)P(x) + B(x)Q(x) \equiv 1$. So obviously, (in any extension field), $P(x)$ and $Q(x)$ can't have common roots.

Proposition: Irreducible polynomials can't have multiple roots.

Let $P(x)$ be an irreducible polynomial over field F. The derivative $P'(x)$ is also a F-polynomial. Note that $\deg P' < \deg P$, $P' \neq 0$. Since P is irreducible, P and P' have no non trivial common factors. Therefore, in $F[x]$, there exists an identity $A(x)P(x) + B(x)P'(x) \equiv 1$. This implies that, (in any extension field), P and P' can't have common roots.

Now, if P has a multiple root α, in a certain extension field, then one can easily see that α is also a root of P'. Since this can't happen, so P has no multiple roots.

Let F be a field, let α be a root of a polynomial over F. α may not be in F. Define $F[\alpha]$ to be the set of all those numbers that can be expressed as polynomials of α with coefficients in F. In other words, $F[\alpha]$ consists of the following kind of numbers, $a_0 + a_1\alpha + a_2\alpha^2 + \cdots$, where there are only finite many entries in the sum and the coefficients $a_0, a_1, a_2, \cdots$ are all in F. If α is a root of a polynomial of degree n, then α^n can expressed as a linear combination of $1, \alpha, \cdots, \alpha^{n-1}$ with coefficients in F. Therefore, all the elements of $F[\alpha]$ can be expressed as linear combinations of $1, \alpha, \cdots, \alpha^{n-1}$ with coefficients in F. Therefore,

Proposition: Let F be a field, α be a root of a polynomial over F, then $F[\alpha]$ is a finite dimensional F-vector space.

Example: $Q[\sqrt{2}]$ consists of the following kind of numbers
$a + b\sqrt{2} + c\sqrt{2}^2 + d\sqrt{2}^3 + \cdots$, where $a, b, c, d, \cdots$ are rational
Since $\sqrt{2}^2 = 2, \sqrt{2}^3 = 2\sqrt{2}, \cdots$,
$Q[\sqrt{2}] = \{a + b\sqrt{2} : a, b \in \mathbb{Q}\}$

Proposition: Assume that F is a field, $P(x)$ is an irreducible F-polynomial of degree n, α is one of its roots. Then $F[\alpha]$ is an n-dimensional vector space over F.

Proof: As mentioned earlier, elements in $F[\alpha]$ can all be expressed as an F-linear combinations of $1, \alpha, \cdots, \alpha^{n-1}$. We now show that $\{1, \alpha, \cdots, \alpha^{n-1}\}$ are linearly independent, thus forming a base for $F[\alpha]$ over F.

Suppose $\{1, \alpha, \cdots, \alpha^{n-1}\}$ are linearly dependent. Then
$a_0 + a_1\alpha + \cdots + a_{n-1}\alpha^{n-1} = 0$, where not all the coefficients are zero.
Let
$Q(x) = a_0 + a_1 x + \cdots + a_{n-1} x^{n-1}$. The polynomials $P(x)$ and $Q(x)$ have a common root α, so must have common factors. Since $P(x)$ is

already irreducible, $P(x)$ must be a factor of $Q(x)$. This is not possible because $P(x)$ has degree n while $Q(x)$ has degree$\leq n-1$. #

Proposition: Let F be a field, α be a root of an F-polynomial. Then $F[\alpha]$ is actually a field.

Proof: Assume that $P(x)$ is an irreducible F-polynomial with α as one of its roots, degree of $P(x)$ is n. Consider any element $\beta \in F[\alpha] - 0$. Let's show that β is invertible in $F[\alpha]$. Write $\beta = a_0 + a_1\alpha + \cdots + a_{n-1}\alpha^{n-1}$. Let $Q(x) = a_0 + a_1 x + \cdots + a_{n-1}x^{n-1}$. Since $P(x)$ is irreducible and $Q(x)$ is a polynomial of a lower degree, they can't have common factors. So there exist F-polynomials A and B, such that $AP + BQ \equiv 1$. Let $x = \alpha$. Then $A(\alpha) \cdot 0 + B(\alpha)\beta = 1$. This proves that $1/\beta$ is still in $F[\alpha]$. #

Let F be a field, $\alpha_1, \cdots, \alpha_n$ be arbitrary numbers. $F[\alpha_1, \cdots, \alpha_n]$ is defined to be the set of all the following kind of finite sums with coefficients in F

$$\sum_{i_1, \cdots, i_n \geq 0} a_{i_1 \cdots i_n} \alpha_1^{i_1} \cdots \alpha_n^{i_n}$$

Note that:
(1) $F[\alpha_1, \cdots, \alpha_n] = F[\alpha_1][\alpha_2] \cdots [\alpha_n]$.
(2) If $\alpha_1, \cdots, \alpha_n$ are all of finite degree over F, then $F[\alpha_1, \cdots, \alpha_n]$ is a field.

Proof: Since α_1 has finite degree over F, $F[\alpha_1]$ is a field. Since α_2 has finite degree over F, it has finite degree over $F[\alpha_1]$, so $F[\alpha_1][\alpha_2]$ is a field. And so on.

If F is a field, E is another field containing F, then E is called an extension of F. We use $F \leq E$ to denote a field extension.

Given fields $F \leq E$, obviously, E can be regarded as a F-vector space. The dimension of this vector space, $\dim_F E$, is also denoted as $[E : F]$.

Proposition: Given fields $A \leq B \leq C$, $[C:A] = [C:B][B:A]$.

Proof: Let $\{\alpha_i\}$ be a linear base for B over A. Let $\{\beta_j\}$ be a linear base for C over B. Let's show that $\{\alpha_i \beta_j\}$ form a linear base for C over A.

Consider any element $z \in C$. Write $z = y_j \beta_j$, $y_j \in B$.

> The symbol $\sum$ omitted here.
> So $y_j \beta_j$ really mean $\sum_j y_j \beta_j$.

Write $y_j = x_{ij} \alpha_i$, $x_{ij} \in A$. Then $z = x_{ij} \alpha_i \beta_j$. This proves that $\{\alpha_i \beta_j\}$ span the whole C over A.

We now show that $\{\alpha_i \beta_j\}$ are linearly independent over A:
$x_{ij} \alpha_i \beta_j = 0$, $x_{ij} \in A \implies$
$(x_{ij} \alpha_i) \beta_j = 0$, $x_{ij} \alpha_i \in B \implies$
$x_{ij} \alpha_i = 0 \implies$
$x_{ij} = 0$. #

6.2 Cubic equations without elementary solutions

A real number is said to have an elementary construction if it can be constructed from rational numbers via $+, -, \times, \div$, plus real radical operations. A "real radical" is a certain $\sqrt[n]{x}$ where x is positive, so that no complex number will arise.

Theorem: Let $P(x)$ be a cubic polynomial with rational coefficients. If it is irreducible with all its roots real, then it has no elementary solution, i.e., none of its three roots has an elementary construction.

Proof:

A real radical extension of a field F is a certain $F[r]$, where $r = \sqrt[n]{R}$, R is a positive number in F. Let's call n the index of the extension. Assume the above theorem isn't true. Then there exists a series of real radical extensions
$$\mathbb{Q} < \mathbb{Q}[r_1] < \mathbb{Q}[r_1][r_2] < \cdots$$
so that eventually, one of the extensions ends up containing one of the roots of $P(x)$. Note that if a, b are positive integers, then $\square^{1/(ab)} = (\square^{1/a})^{1/b}$. Therefore, we may well assume that all the indices in the above real radical extensions are prime numbers. We may further assume F is the field in the above series of extensions such that
   $F = \mathbb{Q}[r_1][r_2] \cdots [r_\square]$ contains none of the three roots
   $F[r]$ contains one of the three roots
   $r = \sqrt[p]{R}$, p is a prime number, R is a positive number in F.

Take a look at polynomial $x^p - R$ over F. Let's explain why it is irreducible: Its roots are $\{re^{i2\pi*/p}, * = 0, \cdots, p-1\}$. If it splits into the product of two non-trivial F-polynomials, then the constant term of either of the two must be certain $(-r)^n e^{i\theta}$, where $1 \leq n \leq p-1$.

This term must be in F, which contains only real numbers. So $e^{i\theta}$ must be ± 1. Therefore r^n must be in F, $1 \le n \le p-1$. Since p is prime, n and p are relatively prime. There must be integers a and b such that $an + bp = 1$. Therefore $r = (r^n)^a \cdot (r^p)^b$ must be in F. Contradiction. In conclusion, the F-polynomial $x^p - R$ is irreducible.

In particular, the dimension of $F[r]$ over F is p.

Note that this implies that there are no intermediate fields between $F < F[r]$: If E is an intermediate field, then $p = [F[r] : F] = [F[r] : E] \times [E : F]$. So either $[E : F]$ or $[F[r] : E]$ must be 1, i.e., E is equal to F or $F[r]$.

Assume α is a root of $P(x)$ such that $\alpha \in F[r]$. Then $F \le F[\alpha] \le F[r]$. So $F[\alpha]$ is either F or $F[r]$. Since $\alpha \notin F$, $F[\alpha] = F[r]$. So $r = Q(\alpha)$, where $Q(x)$ is an F-polynomial. Consider F-polynomial $Q(x)^p - R$, α being one of its roots. On the other hand, the cubic polynomial $P \in \mathbb{Q}[x]$, when regarded as a F-polynomial, is still irreducible, because if it were reducible over F, one of its three roots would have to be in F, contradiction. Two conclusions from this discussion: (1) $[F[\alpha] : F] = 3$, so $p = 3$, anyway p must be an odd prime. (2) Over F, $P(x)$ is irreducible, $P(x)$ and $Q(x)^p - R$ have a common root. Therefore $P(x)$ must be a factor of $Q(x)^p - R$. And therefore, all three roots of $P(x)$ must be roots of $Q(x)^p - R$.

Let the three roots of $P(x)$ be α, β, γ. Then $Q(\alpha), Q(\beta), Q(\gamma)$ are all roots of $x^p - R = 0$.

The roots of $x^p - R = 0$ are $\{re^{i2\pi n/p}, 0 \le n \le p-1\}$. Among these, r is the only real one. The others are all complex numbers. This is because for $e^{i2\pi n/p}$ ($1 \le n \le p-1$) to be real, $2n/p$ must be an integer, $1 \le n \le p-1$. Not possible when $p \ge 3$.

Since $Q(\alpha), Q(\beta), Q(\gamma)$ are all real, they must all be equal to r.
Write $Q = ax^2 + bx + c \in F[x]$. Then
$r = (Q(\alpha) + Q(\beta) + Q(\gamma))/3 = a(\alpha^2 + \beta^2 + \gamma^2)/3 + b(\alpha + \beta + \gamma)/3 + c$
$\alpha^2 + \beta^2 + \gamma^2$, $\alpha + \beta + \gamma$ can all be expressed in terms of the coefficients of P, which are in $\mathbb{Q}$.

We conclude that $r \in F$. Contradiction. This proves the theorem. #

Example: $x^3 - 3x + 1 = 0$ doesn't have an elementary solution.

We first point out that this equation can't have rational roots. Let's look for potential rational roots: Firstly, 0 and ± 1 are not its roots. Suppose n/m is a potential root, where n and m are non-trivial integers, relatively prime.
$$(n/m)^3 - 3(n/m) + 1 = 0$$
$$n^3 - 3nm^2 + m^3 = 0$$
$$n(n^2 - 3m^2) + m^3 = 0$$
Impossible.

We next point out that this polynomial is irreducible over $\mathbb{Q}$: This is a degree 3 polynomial. If it were reducible, there would be rational roots.

By the theorem just proven, this cubic equation can't have an elementary solution, i.e., none of its three roots, $2\sin 10°, 2\sin 50°, -2\sin 70°$, can have an elementary expression.

Corollary: $\sin 10°$ cannot be constructed from rational numbers, via $+, -, \times, \div$, plus real radical operations.

7 Quartic equations

$\underline{x^4 + ax^3 + bx^2 + cx + d = 0}$:

The procedure for solving a quartic equation consists of the following steps.

Let $P(x) = x^4 + ax^3 + bx^2 + cx + d$
Consider equation $P'(\lambda)^2 = (4\lambda + a)^2 P(\lambda)$
This equation on λ turns out to be of degree ≤ 3.
Solve it, choosing one solution λ.

Let $u = \sqrt[4]{P(\lambda)}, v = \lambda$
Note that if $P(\lambda) = 0$, then we already have a root of the original equation, which can then be reduced to an equation of degree 3, and solved. So assume $P(\lambda) \neq 0$.
Write $x = uy + v$

The original equation now reduces to one on y, but with symmetric coefficients. Polynomial equations with symmetric coefficients can be reduced to one with degree half of that of the original.

$\underline{\text{Example: } x^4 = 1 + 4x}$

This equation has two real solutions.
Let's find them.

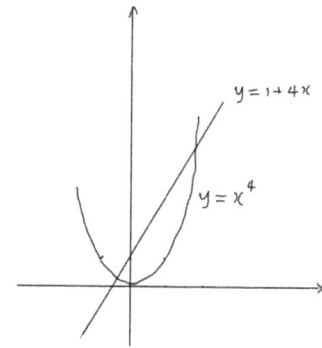

$P(x) = x^4 - 4x - 1$
$(4\lambda^3 - 4)^2 = (4\lambda)^2(\lambda^4 - 4\lambda - 1)$
$\lambda^6 - 2\lambda^3 + 1 = \lambda^2(\lambda^4 - 4\lambda - 1)$
$2\lambda^3 + \lambda^2 + 1 = 0$
$\lambda = -1$
$P(-1) = 4$
$u = \sqrt{2}, v = -1$
$x = \sqrt{2}y - 1$
$(\sqrt{2}y - 1)^4 - 4(\sqrt{2}y - 1) - 1 = 0$
$(\sqrt{2}y)^4 - 4(\sqrt{2}y)^3 + 6(\sqrt{2}y)^2 - 4(\sqrt{2}y) + 1 - 4(\sqrt{2}y - 1) - 1 = 0$
$4y^4 - 8\sqrt{2}y^3 + 12y^2 - 8\sqrt{2}y + 4 = 0$
$\boxed{y^4 - 2\sqrt{2}y^3 + 3y^2 - 2\sqrt{2}y + 1 = 0}$ ⟵ symmetric
$y^2 - 2\sqrt{2}y + 3 - 2\sqrt{2}(1/y) + (1/y^2) = 0$
$(y^2 + 1/y^2) - 2\sqrt{2}(y + 1/y) + 3 = 0$
$(y + 1/y)^2 - 2\sqrt{2}(y + 1/y) + 1 = 0$
Let $z = y + 1/y$
$z^2 - 2\sqrt{2}z + 1 = 0$
$(z - \sqrt{2})^2 = 1$
$z = \sqrt{2} \pm 1$
$y + 1/y = \sqrt{2} \pm 1$
$y^2 + 1 = (\sqrt{2} \pm 1)y$

$$\left(y - \frac{\sqrt{2} \pm 1}{2}\right)^2 = \left(\frac{\sqrt{2} \pm 1}{2}\right)^2 - 1 = \frac{2 \pm 2\sqrt{2} + 1}{4} - 1 = \frac{\pm 2\sqrt{2} - 1}{4}$$

For the real solutions

$$\left(y - \frac{\sqrt{2} + 1}{2}\right)^2 = \frac{2\sqrt{2} - 1}{4}$$

$$y = \frac{\sqrt{2} + 1 \pm \sqrt{2\sqrt{2} - 1}}{2}$$

$$x = \sqrt{2}y - 1 = \frac{\sqrt{2} + 1 \pm \sqrt{2\sqrt{2} - 1}}{\sqrt{2}} - 1$$

The two real roots are

$$x = \frac{1}{\sqrt{2}} \pm \sqrt{\sqrt{2} - \frac{1}{2}}$$

Exercise:

For equation $x^4 + ax^2 + px + q = 0$, show that
$\lambda^3 + 2a\lambda^2 + (a^2 - 4q)\lambda - p^2 = 0$

$$x^4 + ax^2 + px + q = \left[x^2 + \frac{a}{2} + \frac{\lambda}{2} + \left(\sqrt{\lambda}x - \frac{p}{2\sqrt{\lambda}}\right)\right] \times$$
$$\left[x^2 + \frac{a}{2} + \frac{\lambda}{2} - \left(\sqrt{\lambda}x - \frac{p}{2\sqrt{\lambda}}\right)\right]$$

When $a = 0$, show that

$$\frac{\lambda}{2} = \sqrt[3]{\left(\frac{p}{4}\right)^2 + \sqrt{\left(\frac{p}{4}\right)^4 - \left(\frac{q}{3}\right)^3}} + \sqrt[3]{\left(\frac{p}{4}\right)^2 - \sqrt{\left(\frac{p}{4}\right)^4 - \left(\frac{q}{3}\right)^3}}$$

Exercise:

Show that
$x^4 - 4x - 1 = [x^2 + 1 + \sqrt{2}(x+1)][x^2 + 1 - \sqrt{2}(x+1)]$

8 $\cos(2\pi/p)$

$\boxed{\cos(2\pi/5)}$

How to compute $\cos(\pi/5)$

Let $z = e^{\pi i/5}$
$z^5 = -1$, $z^5 + 1 = 0$
$(z+1)(z^4 - z^3 + z^2 - z + 1) = 0$
$z^4 - z^3 + z^2 - z + 1 = 0$
$z^2 - z + 1 - 1/z + 1/z^2 = 0$
$(z^2 + 1/z^2) - (z + 1/z) + 1 = 0$
$(z + 1/z)^2 - (z + 1/z) - 1 = 0$
Let $x = z + 1/z = z + \bar{z} = 2\cos(\pi/5)$
$x^2 - x - 1 = 0$
$x = \dfrac{1 \pm \sqrt{5}}{2} \implies x = \dfrac{1 + \sqrt{5}}{2}$

$\cos(\pi/5) = \dfrac{\sqrt{5} + 1}{4}$

How to compute $\cos(2\pi/5)$

Let $z = e^{2\pi i/5}$
$z^5 = 1$, $z^5 - 1 = 0$
$(z-1)(z^4 + z^3 + z^2 + z + 1) = 0$
$z^4 + z^3 + z^2 + z + 1 = 0$
$z^2 + z + 1 + 1/z + 1/z^2 = 0$
$(z^2 + 1/z^2) + (z + 1/z) + 1 = 0$
$(z + 1/z)^2 + (z + 1/z) - 1 = 0$
Let $x = z + 1/z = z + \bar{z} = 2\cos(2\pi/5)$
$x^2 + x - 1 = 0$

$$x = \frac{-1 \pm \sqrt{5}}{2} \implies x = \frac{-1 + \sqrt{5}}{2}$$

$$\cos(2\pi/5) = \frac{\sqrt{5} - 1}{4}$$

How to draw a triangle to determine $\cos(\pi/5)$ and $\cos(2\pi/5)$

$\pi/5 = 36°$

$$\frac{2}{x} = \frac{x}{2-x} \implies x = \sqrt{5} - 1$$

$$\cos(\pi/5) = \frac{1}{\sqrt{5} - 1} = \frac{\sqrt{5} + 1}{4}$$

$$\cos(2\pi/5) = 2\left(\frac{\sqrt{5} + 1}{4}\right)^2 - 1 = \frac{\sqrt{5} - 1}{4}$$

Exercise: Show that

$$x^4 + x^3 + x^2 + x + 1 = \left(x^2 + \frac{1+\sqrt{5}}{2}x + 1\right)\left(x^2 + \frac{1-\sqrt{5}}{2}x + 1\right)$$

Proof: Let $z = e^{2\pi i/5}$

$$x^4 + x^3 + x^2 + x + 1 = (x-z)(x-z^2)(x-z^3)(x-z^4)$$

$$(x-z)(x-z^4) = x^2 - (z+z^4)x + 1 = x^2 - (z+\bar{z})x + 1$$
$$= x^2 - 2\cos(2\pi/5)x + 1$$

$$(x-z^2)(x-z^3) = x^2 - (z^2+z^3)x + 1 = x^2 - (z^2+\bar{z}^2)x + 1$$
$$= x^2 - 2\cos(4\pi/5)x + 1 = x^2 - 2\cos(\pi/5)x + 1$$

$\ldots$

$\boxed{\cos(2\pi/7)}$

Proposition 1:

The roots of the equation $x^3 + x^2 - 2x - 1 = 0$ are $2\cos(2\pi/7)$, $-2\cos(3\pi/7)$, $-2\cos(\pi/7)$.

Proof: Let $z = e^{2\pi i/7}$

$$z^7 = 1$$
$$z^7 - 1 = 0$$
$$(z-1)(z^6 + z^5 + z^4 + z^3 + z^2 + z + 1) = 0$$
$$z^6 + z^5 + z^4 + z^3 + z^2 + z + 1 = 0$$
$$z^3 + z^2 + z + 1 + 1/z + 1/z^2 + 1/z^3 = 0$$
$$(z^3 + 1/z^3) + (z^2 + 1/z^2) + (z + 1/z) + 1 = 0$$
$$(z + 1/z)^3 = (z^3 + 1/z^3) + 3(z + 1/z)$$
$$(z + 1/z)^2 = (z^2 + 1/z^2) + 2$$
$$(z + 1/z)^3 - 3(z + 1/z) + (z + 1/z)^2 - 2 + (z + 1/z) + 1 = 0$$
$$(z + 1/z)^3 + (z + 1/z)^2 - 2(z + 1/z) - 1 = 0$$
Let $x = z + 1/z$
$$x^3 + x^2 - 2x - 1 = 0$$

Note that actually z can be any of $e^{2n\pi i/7}$, $1 \leq n \leq 6$, with $x = 2\cos(2n\pi/7)$. Therefore the roots of $x^3 + x^2 - 2x - 1 = 0$ are

$$2\cos(2\pi/7)$$
$$2\cos(4\pi/7) = -2\cos(3\pi/7)$$
$$2\cos(6\pi/7) = -2\cos(\pi/7). \quad \#$$

Exercise: Show that

$$\cos(\pi/7) - \cos(2\pi/7) + \cos(3\pi/7) = 1/2$$
$$\cos(\pi/7) \cdot \cos(2\pi/7) \cdot \cos(3\pi/7) = 1/8$$
$$1/\cos(\pi/7) - 1/\cos(2\pi/7) + 1/\cos(3\pi/7) = 4$$

Proposition 2: $\cos(\pi/7)$ cannot be constructed from rational numbers, via $+, -, \times, \div$, plus real radical operations.

Proof: The equation $x^3 + x^2 - 2x - 1 = 0$ has no rational roots, and thus is irreducible: This is because the only potential rational roots are ± 1, but neither really is. The claim then follows from the theorem in §6.2.

Proposition 3:

(i) $x^6 + x^5 + x^4 + x^3 + x^2 + x + 1 = \left(x^3 + \dfrac{1-\sqrt{7}i}{2}x^2 - \dfrac{1+\sqrt{7}i}{2}x - 1 \right) \times$
$$\left(x^3 + \dfrac{1+\sqrt{7}i}{2}x^2 - \dfrac{1-\sqrt{7}i}{2}x - 1 \right)$$

(ii) $\sin(2\pi/7) + \sin(3\pi/7) - \sin(\pi/7) = \sqrt{7}/2$

Proof:
Let $z = e^{2\pi i/7}$
$$x^6 + x^5 + x^4 + x^3 + x^2 + x + 1 = (x-z)(x-z^2)(x-z^3)(x-z^4)(x-z^5)(x-z^6)$$

We divide the roots $\{z, z^2, z^3, \cdots\}$ into two groups. The first group consists of $\{z, z^2, z^4, [z^8], \cdots\}$. The rest of the roots form the other group.

Let $\alpha = z + z^2 + z^4$, $\beta = z^3 + z^5 + z^6$
$\alpha + \beta = z + \cdots + z^6 = -1$
$\alpha\beta = z^4 + z^6 + z^7 + z^5 + z^7 + z^8 + z^7 + z^9 + z^{10}$
$= z^4 + z^6 + 1 + z^5 + 1 + z + 1 + z^2 + z^3$
$= 2$

$(\alpha - \beta)^2 = (\alpha + \beta)^2 - 4\alpha\beta = 1 - 8 = -7$
$\alpha - \beta = \pm\sqrt{7}i$

$$\alpha = \frac{-1 \pm \sqrt{7}i}{2}, \quad \beta = \frac{-1 \mp \sqrt{7}i}{2}$$

$$z + z^2 + z^4 = \frac{-1 \pm \sqrt{7}i}{2}$$

$\sin(2\pi/7) + \sin(4\pi/7) + \sin(8\pi/7) = \pm\sqrt{7}/2$
$\sin(2\pi/7) + \sin(3\pi/7) - \sin(\pi/7) = \pm\sqrt{7}/2$
The expression on the left is > 0, therefore
$\sin(2\pi/7) + \sin(3\pi/7) - \sin(\pi/7) = \sqrt{7}/2$

$$\alpha = \frac{-1 + \sqrt{7}i}{2}, \quad \beta = \frac{-1 - \sqrt{7}i}{2}$$

$(x - z)(x - z^2)(x - z^4) = x^3 + ax^2 + bx + c$

$$a = -(z + z^2 + z^4) = \frac{1 - \sqrt{7}i}{2}$$

$$b = zz^2 + zz^4 + z^2z^4 = z^3 + z^5 + z^6 = \frac{-1 - \sqrt{7}i}{2}$$

$c = -zz^2z^4 = -1$

$(x - z^3)(x - z^5)(x - z^6)$
$= (x - \bar{z}^4)(x - \bar{z}^2)(x - \bar{z})$
$= \overline{(x - z)(x - z^2)(x - z^4)}$
$= x^3 + \bar{a}x^2 + \bar{b}x + \bar{c}.$ #

About $\cos(2\pi/p)$ in general

A general version of what we proved in §6.2 leads to the following theorem: Given a prime p, $\cos(2\pi/p)$ can be constructed from rational numbers via $+, -, \times, \div$ plus real radical operations if and only p is of type $1 + 2^*$.

The first few primes:

> For $p = 5$, $5 - 1 = 4 = 2^2$, so $\cos(2\pi/5)$ can be constructed via elementary means. We have shown that $\cos(2\pi/5) = (\sqrt{5}-1)/4$.
>
> For $p = 7$, $7 - 1 = 6 = 2 \times 3$, so $\cos(2\pi/7)$ cannot be constructed via elementary means.
>
> For $p = 11$, $11 - 1 = 10 = 2 \times 5$, so $\cos(2\pi/11)$ cannot be constructed via elementary means.
>
> For $p = 13$, $13 - 1 = 12 = 2^2 \times 3$, so $\cos(2\pi/13)$ cannot be constructed via elementary means.
>
> For $p = 17$, $17 - 1 = 16 = 2^4$, so $\cos(2\pi/17)$ can be constructed via elementary means.
>
> Note that the next prime of type $p = 1 + 2^*$ is $p = 257$, $257 - 1 = 256 = 2^8$.

$\cos(2\pi/17)$

Since $\cos(2\pi/17)$ can be expressed as a construction from rational numbers via $+, -, \times, \div$ plus real radical operations, we want to know what that expression looks like.

Step (1)

Let $z = e^{2\pi i/17}$
$z^{17} = 1$
$(z - 1)(1 + z + \cdots + z^{16}) = 0$

$1 + z + \cdots + z^{16} = 0$
$z + \cdots + z^{16} = -1$

We divide $\{z, \cdots, z^{16}\}$ into two groups. The first group consists of $\{z, z^2, z^4, z^8, z^{16}, \cdots\}$. The rest form the other group. Note that,
$z^{32} = z^{32-17} = z^{15}$
$z^{64} = z^{2 \times 15} = z^{30} = z^{30-17} = z^{13}$
$z^{128} = z^{2 \times 13} = z^{26} = z^{26-17} = z^9$
$\left(z^{256} = z^{2 \times 9} = z^{18} = z^{18-17} = z\right)$
$\cdots$

Group 1: $z, z^2, z^4, z^8, z^9, z^{13}, z^{15}, z^{16}$
Group 2: $z^3, z^5, z^6, z^7, z^{10}, z^{11}, z^{12}, z^{14}$

Let $x = z + z^2 + z^4 + z^8 + z^9 + z^{13} + z^{15} + z^{16}$
$y = z^3 + z^5 + z^6 + z^7 + z^{10} + z^{11} + z^{12} + z^{14}$

$x + y = -1$

$xy = z^4 + z^6 + z^7 + z^8 + z^{11} + z^{12} + z^{13} + z^{15}$
$ z^5 + z^7 + z^8 + z^9 + z^{12} + z^{13} + z^{14} + z^{16}$
$ z^7 + z^9 + z^{10} + z^{11} + z^{14} + z^{15} + z^{16} + z$
$ z^{11} + z^{13} + z^{14} + z^{15} + z + z^2 + z^3 + z^5$
$ z^{12} + z^{14} + z^{15} + z^{16} + z^2 + z^3 + z^4 + z^6$
$ z^{16} + z + z^2 + z^3 + z^6 + z^7 + z^8 + z^{10}$
$ z + z^3 + z^4 + z^5 + z^8 + z^9 + z^{10} + z^{12}$
$ z^2 + z^4 + z^5 + z^6 + z^9 + z^{10} + z^{11} + z^{13}$
$= 4(z + \cdots + z^{16}) = -4$

$x, y = \dfrac{-1 \pm \sqrt{17}}{2}$

$z + z^2 + z^4 + z^8 + z^9 + z^{13} + z^{15} + z^{16} = \dfrac{\sqrt{17} - 1}{2}$

$z^3 + z^5 + z^6 + z^7 + z^{10} + z^{11} + z^{12} + z^{14} = -\dfrac{\sqrt{17} + 1}{2}$

(Use a calculator to determine which is which)

Step (2)

Now, further divide $\{z, z^2, z^4, z^8, z^9, z^{13}, z^{15}, z^{16}\}$ into two groups. The first group consists of $\{z, z^4, z^{16}, \cdots\}$. The rest form the other group. Note that, $z^{64} = z^{13}$, $z^{256} = z$. So the two groups are
$\{z, z^4, z^{13}, z^{16}\}$
$\{z^2, z^8, z^9, z^{15}\}$

Let $x = z + z^4 + z^{13} + z^{16}$
$\quad y = z^2 + z^8 + z^9 + z^{15}$

$x + y = \dfrac{\sqrt{17} - 1}{2}$

$xy = z^3 + z^9 + z^{10} + z^{16}$
$\quad\quad z^6 + z^{12} + z^{13} + z^2$
$\quad\quad z^{15} + z^4 + z^5 + z^{11}$
$\quad\quad z + z^7 + z^8 + z^{14}$
$\quad = z + \cdots + z^{16}$
$\quad = -1$

$x, y = \dfrac{\sqrt{17} - 1}{4} \pm \sqrt{1 + \left(\dfrac{\sqrt{17} - 1}{4}\right)^2}$

Denote $a = \dfrac{\sqrt{17} - 1}{4} + \sqrt{1 + \left(\dfrac{\sqrt{17} - 1}{4}\right)^2}$

$\tilde{a} = \dfrac{\sqrt{17} - 1}{4} - \sqrt{1 + \left(\dfrac{\sqrt{17} - 1}{4}\right)^2}$

$z + z^4 + z^{13} + z^{16} = a$
$z^2 + z^8 + z^9 + z^{15} = \tilde{a}$
(Use a calculator to determine which is which)

Step (3)

Lastly, we divide $\{z, z^4, z^{13}, z^{16}\}$ into $\{z, z^{16}\}$ and $\{z^4, z^{13}\}$.
But note that
$(z + z^{16}) + (z^4 + z^{13}) = a$
$(z + z^{16}) \times (z^4 + z^{13}) = z^5 + z^{14} + z^3 + z^{12} = z^3 + z^5 + z^{12} + z^{14}$
We need to determine $z^3 + z^5 + z^{12} + z^{14}$ first.

$$z^3 + z^5 + z^6 + z^7 + z^{10} + z^{11} + z^{12} + z^{14} = -\frac{\sqrt{17}+1}{2}$$

Let $x = z^3 + z^5 + z^{12} + z^{14}$
$y = z^6 + z^7 + z^{10} + z^{11}$

$x + y = -\dfrac{\sqrt{17}+1}{2}$

$\begin{aligned}xy =& z^9 + z^{10} + z^{13} + z^{14} \\ & z^{11} + z^{12} + z^{15} + z^{16} \\ & z + z^2 + z^5 + z^6 \\ & z^3 + z^4 + z^7 + z^8 \\ =& z + \cdots + z^{16} \\ =& -1\end{aligned}$

$x, y = -\dfrac{\sqrt{17}+1}{4} \pm \sqrt{1 + \left(\dfrac{\sqrt{17}+1}{4}\right)^2}$

Denote $b = -\dfrac{\sqrt{17}+1}{4} + \sqrt{1 + \left(\dfrac{\sqrt{17}+1}{4}\right)^2}$

$\tilde{b} = -\dfrac{\sqrt{17}+1}{4} - \sqrt{1 + \left(\dfrac{\sqrt{17}+1}{4}\right)^2}$

$z^3 + z^5 + z^{12} + z^{14} = b$
$z^6 + z^7 + z^{10} + z^{11} = \tilde{b}$
(Use a calculator to determine which is which)

Step (4)

So divide $\{z, z^4, z^{13}, z^{16}\}$ into $\{z, z^{16}\}$ and $\{z^4, z^{13}\}$.

Let $x = z + z^{16}, y = z^4 + z^{13}$
$x + y = a, xy = b$
$x, y = a/2 \pm \sqrt{(a/2)^2 - b}$

$z + z^{16} = a/2 + \sqrt{(a/2)^2 - b}$
$z^4 + z^{13} = a/2 - \sqrt{(a/2)^2 - b}$
(Use a calculator to determine which is which)

The first identity implies
$z + z^{-1} = a/2 + \sqrt{(a/2)^2 - b}$
$2\cos(2\pi/17) = a/2 + \sqrt{(a/2)^2 - b}$

Step (5)

Let's simplify a^2
$$\begin{aligned} a^2 &= (z + z^4 + z^{13} + z^{16})^2 \\ &= z^2 + z^5 + z^{14} + 1 \\ &\quad z^5 + z^8 + 1 + z^3 \\ &\quad z^{14} + 1 + z^9 + z^{12} \\ &\quad 1 + z^3 + z^{12} + z^{15} \\ &= 4 + 2(z^3 + z^5 + z^{12} + z^{14}) + (z^2 + z^8 + z^9 + z^{15}) \\ &= 4 + 2b + \tilde{a} \end{aligned}$$

Step (6)

Exercise: Complete the final step to verify that

$$\cos(2\pi/17) = \frac{-1+\sqrt{17}+\sqrt{34-2\sqrt{17}}+2\sqrt{17+3\sqrt{17}-\sqrt{34-2\sqrt{17}}-2\sqrt{34+2\sqrt{17}}}}{16}$$

9 Elementary properties of the prime numbers

9.1 There are infinite many primes

The ancient Greek argument demonstrating that there are infinite many primes is as follows: Assume that there are only finitely many primes. Let n be their product. Then $n+1$ cannot contain any prime. Contradiction.

Alternatively, the following identity also implies that there exist infinite many primes:

$$\prod \frac{1}{1-1/p} = \sum 1/n = +\infty$$

Note that here, $\sum 1/n$ denotes the sum of all $1/n$ with n enumerates all positive integers, $\prod \frac{1}{1-1/p}$ denotes the product of all $\frac{1}{1-1/p}$ with p enumerating all the primes. The reason for this identity:

$$\prod \frac{1}{1-1/p} = \prod (1+1/p+1/p^2+\cdots) = \text{(as the product expands)} = \sum 1/n$$

Exercise: $\sum 1/p = +\infty$

How to prove it:

$$\prod \frac{1}{1-1/p} = \sum 1/n = +\infty$$

$$\ln \prod \frac{1}{1-1/p} = +\infty$$

$$\sum \ln \frac{1}{1-1/p} = +\infty$$

Lemma: $\ln \frac{1}{1-x} \leq 2x, 0 \leq x \leq 1/2$
$\ln \frac{1}{1-x} \leq 2x \Longleftarrow$
$2x + \ln(1-x) \geq 0$
Let $f(x) = 2x + \ln(1-x)$
$f(0) = 0$
$f'(x) = 2 - \frac{1}{1-x} \geq 0, 0 \leq x \leq 1/2$

Since $\sum \ln \frac{1}{1-1/p} \leq 2 \sum 1/p$
Therefore $\sum 1/p = +\infty$.

9.2 The field $\mathbb{Z}_p$

In mathematics, $\mathbb{Z}$ denotes the set of all integers. Given a prime p, $\mathbb{Z}_p$ is still the set of all integers, but with a new understanding, which is that, given two integers n and m, if their difference can be written as $n - m = p \cdot \square$, where "$\square$" is integer, then n and m are considered the same in $\mathbb{Z}_p$. Thus, unlike $\mathbb{Z}$ which contains infinite many elements, $\mathbb{Z}_p$ contains only a finite number of elements:

$$\mathbb{Z}_p = \{0, 1, \cdots, p-1\}$$

In $\mathbb{Z}_p$, the arithmetic operations, $+, -, \times, \div$, all still work "as usual".

Example:
$\mathbb{Z}_5 = \{0, 1, 2, 3, 4\}$
In $\mathbb{Z}_5$,
$3 + 4 = 7 = 7 - 5 = 2$
$3 - 4 = -1 = 5 - 1 = 4$
$3 * 4 = 12 = 12 - 5 * 2 = 2$
$3/4 = 2$, because, $2 * 4 = 8 = 8 - 5 = 3$.

Remark: To see why division still works in $\mathbb{Z}_p$, we just need to explain why $1/n$, where n is a non-zero value in $\mathbb{Z}_p$, exists in $\mathbb{Z}_p$. So $2 \leq n \leq p-1$. Since n and p are relatively prime, there exists identity

$an + bp = 1$, where a and b are integers. In $\mathbb{Z}_p$, this identity becomes $an = 1$, so $1/n = a$.

9.3 Fermat's Little Theorem

Let p be a prime
Let's look at the binomial expansion formula
$$(a+b)^p = a^p + pa^{p-1}b + \frac{p(p-1)}{2}a^{p-2}b^2 + \cdots + pab^{p-1} + b^p$$
The coefficients are $\frac{p(p-1)\cdots(p-i+1)}{i!}, 1 \leq i \leq p-1$. In any of these coefficients, the prime factor p in the numerator will never be cancelled out, because the denominator $i! = 1 \cdot 2 \cdots i$ doesn't have p as one of its factors. In other words, all the coefficients have p as one of its prime factors. Therefore, for arbitrary integers a and b, $(a+b)^p = a^p + b^p + p \cdot \square$, where "$\square$" is integer.

In $\mathbb{Z}_p$, the above identity becomes
$$(a+b)^p = a^p + b^p$$
This implies that, for any $n \in \mathbb{Z}_p$,
$n^p = (1 + 1 + \cdots + 1)^p = 1^p + 1^p + \cdots + 1^p = 1 + 1 + \cdots + 1 = n$

Fermat's Little Theorem:
Let p be a prime. For any $n \in \mathbb{Z}_p - 0$,
$$n^{p-1} = 1$$

Over $\mathbb{Z}_p$, the polynomial $x^{p-1} - 1$ is of degree=$p-1$. The above theorem tells that all the non-zero elements of $\mathbb{Z}_p$ are its roots. And there are exactly $p-1$ many. Therefore there is the following (powerful) decomposition formula
$$x^{p-1} - 1 = (x-1)(x-2) \cdots (x-(p-1))$$

Application: Let p be a prime. Then in $\mathbb{Z}_p$, $(p-2)! = 1$.

To deduce it, let's assume $p \geq 3$.
The decomposition formula for $x^{p-1} - 1$ implies
$(-1)(-2) \cdots (-(p-1)) = -1$.
Note that $p - 1 = -1$.
So $(-1)(-2) \cdots (-(p-2)) = -1$.
$(-1)^{p-2}(p-2)! = -1$.
$(p-2)! = 1$. #

How to prove $(p-2)! = 1$ directly:

In $\mathbb{Z}_p$, all the elements except 0 are invertible. Note that over $\mathbb{Z}_p$, the equation $x^2 = 1$ has just two solutions, $x = 1$ and $x = -1 = p - 1$. So for any of the following elements, $x^{-1} \neq x$:

$$2, 3, \cdots, p-2$$

Let's group the above list of elements into pairs, each pair consisting of certain $\{n, 1/n\}$ where $2 \leq n \leq p-2$. We thus see that $2 \times 3 \times \cdots \times (p-2) = 1$.

9.4 There are infinite many primes in $4\mathbb{Z} + 1$, $4\mathbb{Z} - 1$

Basically we are claiming that there are infinite many primes that can be written as $p = 4n + 1$, and there are also infinite many primes that can be written as $p = 4n - 1$.

There are infinite many primes in $4\mathbb{Z} - 1$:

Proof: Assume there are only finitely many primes in $4\mathbb{Z} - 1$. Let n be the product of all of them.

> When we say a number "contains" a prime, we mean the number has the prime as one of its factors. For example, $150 = 2 \cdot 3 \cdot 5^2$, so 150 contains primes 2, 3, and 5.

Consider $4n-1$. We first point out that it cannot contain any prime of type $4\mathbb{Z}-1$. For if p were a prime of type $4\mathbb{Z}-1$ that divides $4n-1$, then p would divide both $4n-1$ and n, something impossible. Also note that $4n-1$ is odd, so cannot contain 2. Therefore, all the primes in $4n-1$ must be of type $4\mathbb{Z}+1$.

But then, $4n-1$, being the product of a set of numbers in $4\mathbb{Z}+1$, must also be in $4\mathbb{Z}+1$. In other words, $4n-1 = 4m+1$. Again impossible.
#

There are infinite many primes in $4\mathbb{Z}+1$:

Proof: Assume there are only finite many primes of type $4\mathbb{Z}+1$. Let n be their product. Consider $4n^2+1$. Note that it cannot contain any prime of type $4\mathbb{Z}+1$. For if p were a prime of type $4\mathbb{Z}+1$ that divides $4n^2+1$, then p would divide both $4n^2+1$ and n, impossible. So all the primes in $4n^2+1$ are of type $4\mathbb{Z}-1$. But $4n^2+1 = (2n)^2+1$.

> Lemma: For any integer a, a^2+1 contains no prime of type $4\mathbb{Z}-1$.
> Proof: Consider any prime $p \in 4\mathbb{Z}-1$. Let's show that, over $\mathbb{Z}_p$, the equation $x^2+1=0$ has no solution. Assume there is, i.e., there exists $x \in \mathbb{Z}_p$ such that $x^2 = -1$. Write $p = 4m-1$. $(x^2)^{2m-1} = (-1)^{2m-1} = -1$. $x^{4m-2} = -1$. $x^{p-1} = -1$. But Fermat's Little Theorem says $x^{p-1} = 1$. So $1 = -1$, i.e., $p = 2$. Impossible.

So $(2n)^2+1$ contains no prime of type $4\mathbb{Z}-1$. Contradiction. #

10 The integer equation $x^2 + y^2 = z^2$

10.1 Complex integers

A complex number $z = x + iy$, where x and y are integers, is called a complex integer, also known as a Gauss integer. The set of all complex integers is denoted $\mathbb{Z}[i]$. For any $z = x + iy \in \mathbb{Z}[i]$, its norm is defined to be
$$N(z) = x^2 + y^2$$
Note that $N(z)$ is a positive integer, except that $N(0) = 0$.

$N(zw) = N(z)N(w)$:

This is because
$N(zw) = (zw)\overline{(zw)} = zw\bar{z}\bar{w} = z\bar{z}w\bar{w} = N(z)N(w)$

Recall that, given a complex number $z = x + iy$, $\bar{z} = x - iy$.

Units are those elements $z \in \mathbb{Z}[i]$ such that z^{-1} still in $\mathbb{Z}[i]$.

We first point out that $z \in \mathbb{Z}[i]$ is a unit if and only if $N(z) = 1$: If z is a unit, then z^{-1} is also in $\mathbb{Z}[i]$. $zz^{-1} = 1 \implies N(z)N(z^{-1}) = N(1) = 1$. Note that both $N(z)$ and $N(z^{-1})$ are positive integers. So $N(z) = N(z^{-1}) = 1$. On the other hand, if z is a complex integer such that $N(z) = 1$, then $z\bar{z} = 1$, then $z^{-1} = \bar{z}$ is in $\mathbb{Z}[i]$.

The only integer solutions to equation $x^2 + y^2 = 1$ are $(0, \pm 1), (\pm 1, 0)$. Thus the only units in $\mathbb{Z}[i]$ are: $\pm 1, \pm i$

The concept of primes in $\mathbb{Z}[i]$: A complex integer is said to be prime if it cannot be further decomposed into non-trivial products of other complex integers.

First note that, if z is a complex integer such that $N(z)$ is a prime number, then z must be prime in $\mathbb{Z}[i]$: If $z = uv$, then $N(z) = N(u)N(v)$. Since $N(z)$ is a prime number, either $N(u)$ or $N(v)$ must be 1. That is, either u or v must a unit, i.e., must be one of the four trivial elements $\{\pm 1, \pm i\}$.

From this principle we can identify lots of primes in $\mathbb{Z}[i]$. For example $1 \pm i$ are both primes, because $N(1 \pm i) = 2$ is a prime number. $1 \pm 2i$ are also primes, because $N(1 \pm 2i) = 5$ is a prime number.

Conversely, 2 is not a prime in $\mathbb{Z}[i]$, because it can be further decomposed, $2 = (1+i)(1-i)$.

Similarly, 5 is not a prime in $\mathbb{Z}[i]$, because $5 = (1+2i)(1-2i)$, or, $5 = (2+i)(2-i)$. Note that here we appear to have identified two different decompositions of 5 within $\mathbb{Z}[i]$. But actually, these two decompositions are essentially the same: $1 + 2i = i(2-i)$, $1 - 2i = -i(2+i)$. $\pm i$ are units.

3 is a prime in $\mathbb{Z}[i]$: Suppose 3 can be further decomposed into a non-trivial product, $3 = zw$, where both z and w are non-units. Then $N(z)N(w) = N(3) = 9$. Then $N(z) = N(w) = 3$. Write $z = x + iy$. Then $x^2 + y^2 = 3$. This equation has no integer solution.

An Euclid algorithm in $\mathbb{Z}[i]$

In arithmetics, the Euclid algorithm describes what happens when you attempt to divide one integer by another: If a and b are integers, then a/b can be expressed as $a/b = c + d/b$, where c and d are integers, with $|d| < |b|$.

Proposition: There exists an Euclid algorithm in $\mathbb{Z}[i]$ in the sense that, given any two complex integers z and w, z/w can be expressed as $z/w = u + v/w$, where u and v are complex integers, with $N(v) < N(w)$.

The proof can be found after the following lemma:

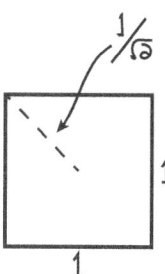

Lemma: For any complex number z, there exists a complex integer z_0 such that $|z - z_0| \leq 1/\sqrt{2}$.

Proof of proposition: Let z and $w \neq 0$ be complex integers. Consider z/w. If z/w is complex integer, we are done. Suppose z/w is not. By the above lemma, there exists a complex integer u such that $|z/w - u| \leq 1/\sqrt{2} < 1$. Write $z/w - u = v/w$, i.e., $z/w = u + v/w$, where v is complex integer.
$|v/w| < 1 \Longrightarrow |v| < |w| \Longrightarrow |v|^2 < |w|^2$, i.e., $N(v) < N(w)$. #

Importance of having an Euclid algorithm in $\mathbb{Z}[i]$

The existence of an Euclid algorithm ensures that primes in $\mathbb{Z}[i]$ behave just like the ordinary primes in integers:

Similarity 1: If complex integers z_1 and z_2 are relatively prime, i.e., don't have non-trivial common factors, then there exist complex integers w_1 and w_2, such that $w_1 z_1 + w_2 z_2 = 1$.

Proof: Let's assume $N(z_1) \geq N(z_2)$. Try dividing z_1 by z_2, so that $z_1/z_2 = q + z_3/z_2$, where $N(z_3) < N(z_2)$. $z_1 = qz_2 + z_3$ implies that z_2 and z_3 are relatively prime. A non-trivial common factor for z_2 and z_3 would be a non-trivial common factor for z_1 and z_2. By introducing mathematical induction on the norm, we have
$$uz_2 + vz_3 = 1$$
$$uz_2 + v(z_1 - qz_2) = 1$$
$$vz_1 + (u - vq)z_2 = 1$$

Similarity 2: If z is a prime that divides $z_1 z_2$, then z must divide either z_1 or z_2.

Proof: If z divides neither z_1 nor z_2, then z is relatively prime to z_1, as well as to z_2. So there exist $uz + u_1 z_1 = 1$, $vz + v_2 z_2 = 1$.
$$(uz + u_1 z_1)(vz + v_2 z_2) = 1$$
$$(uvz + uv_2 z_2 + u_1 v z_1) z + (u_1 v_2) z_1 z_2 = 1$$
So z and $z_1 z_2$ must be relatively prime. Contradiction.

Similarity 3: Any complex integer can be decomposed into the product of its prime factors, and the decomposition is essentially unique.

10.2 The integer equation $x^2 + y^2 = p$

Lemma: Let $p \geq 3$ be a prime number. There exists an integer n such that $n^2 + 1$ contains p as a factor if and only if p is of type $4\mathbb{Z} + 1$.

Proof: Fermat's little theorem says that the elements of $\mathbb{Z}_p - 0$ are precisely the $(p-1)$ roots of the equation $x^{p-1} - 1 = 0$ over $\mathbb{Z}_p$. Suppose $p = 4m + 1$. Then
$x^{p-1} - 1 = x^{4m} - 1 = (x-1)(x+1)(x^2+1)\cdots$
So the equation $x^2 + 1 = 0$ must have two solutions in $\mathbb{Z}_p - 0$. Let n be one of them. Then $n^2 + 1 = 0$ in $\mathbb{Z}_p$, which means $n^2 + 1$ contains p as a prime factor.

On the other hand, if p is not of type $4\mathbb{Z} + 1$, then it is of type $4\mathbb{Z} - 1$. So $p = 4m - 1$. Then $x^{p-1} = x^{4m-2} = (x^2)^{2m-1}$. So
$(x^2)^{2m-1} = 1$, for any $x \in \mathbb{Z}_p - 0$
Since the index $2m - 1$ is odd, x^2 can never be -1. So $x^2 + 1 = 0$ doesn't have any solution over $\mathbb{Z}_p$. That is, for any integer n, $n^2 + 1$ doesn't contain p as a factor.

Theorem: Let $p \geq 3$ be a prime. The integer equation $x^2 + y^2 = p$ has solution if and only if p is of type $4\mathbb{Z} + 1$. When there is solution, the solution is essentially unique.

Proof: Suppose there are integers x, y such that $x^2 + y^2 = p$. Then for the pair $\{x, y\}$, one must be even while the other be odd. Therefore $p = (2a)^2 + (2b+1)^2 = 4a^2 + 4b^2 + 4b + 1 \in 4\mathbb{Z} + 1$

Now suppose p is of type $4\mathbb{Z}+1$. We first point out that p, if considered as a complex integer, is not prime, i.e., it can be further decomposed in $\mathbb{Z}[i]$: The lemma above says that there exists an integer n such that $n^2 + 1$ contains p as a factor. Note that $n^2 + 1 = (n+i)(n-i)$. If p were prime in $\mathbb{Z}[i]$, then, since it divides $(n+i)(n-i)$, it would have to divide either $n+i$ or $n-i$. Say it divides $n+i$. Then $n+i = p(x+iy)$, where x, y are integers. Then $py = 1$, contradiction.

So p is not prime in $\mathbb{Z}[i]$. Say $p = zw$, where $z, w \in \mathbb{Z}[i]$ are not units. $N(z)N(w) = N(p) = p^2$. $N(z)$ and $N(w)$ are not 1, so $N(z) = N(w) = p$. Write $z = x + iy$, then $x^2 + y^2 = p$.

We now explain why the solution is essentially unique: $x^2 + y^2 = p$ implies $(x+iy)(x-iy) = p$. Since $N(x \pm iy) = p$ is a prime number, both $x \pm iy$ are primes in $\mathbb{Z}[i]$. Therefore $p = (x+iy)(x-iy)$ is the decomposition of p into the product its prime factors. This decomposition is essentially unique. #

Identifying all the primes of $\mathbb{Z}[i]$

Theorem:
Let p be a prime number.
The prime decomposition of $p \in \mathbb{Z}[i]$:
   If $p = 2$: $2 = (1+i)(1-i)$
   If $p \in 4\mathbb{Z} + 1$: $p = (x+iy)(x-iy)$, $x^2 + y^2 = p$
   If $p \in 4\mathbb{Z} - 1$: $p \in \mathbb{Z}[i]$ is prime.

The primes described in these decompositions constitute all the primes there are in $\mathbb{Z}[i]$.

Proof that if $p \in 4\mathbb{Z} - 1$, then p is prime in $\mathbb{Z}[i]$: If $p \in \mathbb{Z}[i]$ is not prime, then $p = zw$, where z, w are not units. $N(z)N(w) = p^2$, $N(z), N(w)$ are not 1. So $N(z) = N(w) = p$. Write $z = x + iy$, then $x^2 + y^2 = p$. $\{x, y\}$ must be one even and one odd, which implies $p \in 4\mathbb{Z}+1$, contradiction.

Proof that primes listed in the decompositions are all the primes there are in $\mathbb{Z}[i]$: Consider an arbitrary prime $x + iy \in \mathbb{Z}[i]$. Note that $(x+iy)(x-iy)$ is a positive integer. This positive integer is the product of a few prime numbers. The prime numbers can be further decomposed into primes in $\mathbb{Z}[i]$ as listed above. Therefore $(x + iy)(x - iy)$ can be expressed as a product of the $\mathbb{Z}[i]$-primes listed above. Since $x + iy$ is itself a prime, it must be equal to a unit times one of the primes mentioned in the list above.

10.3 The integer equation $x^2 + y^2 = z^2$

The obvious solutions: Given any two integers (a, b), $(x, y, z) = (a^2 - b^2, 2ab, a^2 + b^2)$ obviously is a solution of $x^2 + y^2 = z^2$. If (x, y, z) is a solution, then you can exchange the first two values, or replace any of the three values by its negative. Also, if (x, y, z) is a solution, then (nx, ny, nz) is also a solution.

Examples:

$(a, b) = (2, 1)$ gives $3^2 + 4^2 = 5^2$
$(a, b) = (3, 2)$ gives $5^2 + 12^2 = 13^2$
$(a, b) = (4, 3)$ gives $7^2 + 24^2 = 25^2$
$(a, b) = (4, 1)$ gives $8^2 + 15^2 = 17^2$
...

Theorem: The integer equation $x^2 + y^2 = z^2$ has no other solutions besides the obvious ones.

Proof: Assume that $\{x, y, z\}$ are integers, relatively prime, such that $x^2 + y^2 = z^2$. We first point out that z cannot have prime factors of type $4\mathbb{Z} - 1$: Suppose $p \in 4\mathbb{Z} - 1$ is a prime factor of z. Note that p remains prime in $\mathbb{Z}[i]$. Since $(x + iy)(x - iy) = z^2$, we conclude that p has to divide either $x + iy$ or $x - iy$. This implies that p divides both x and y. Contradiction.

So if p is a prime factor of z, then $p = 2$, or $p \in 4\mathbb{Z} + 1$. Either way, in $\mathbb{Z}[i]$, there is always decomposition $p = u\bar{u}$, where u, and its conjugate $\bar{u}$, are (different) primes. Since $(x + iy)(x - iy) = z^2$, $(x + iy)(x - iy)$

must contain $(u\bar{u})^2$ in its prime decomposition. Note that $x+iy$ cannot contain both u and $\bar{u}$ as its prime factors, otherwise it would contain $p = u\bar{u}$ as a factor. Then p would divide both x and y, contradiction. The same for $x - iy$. In other words, either $x + iy$ contains u^2 and $x - iy$ contains $\bar{u}^2$, or $x + iy$ contains $\bar{u}^2$ and $x - iy$ contains u^2. By going through all the prime factors of z, we conclude that $x + iy = v^2$, where $v \in \mathbb{Z}[i]$, up to a possible multiplication by i. Write $v = a + ib$. $(a + ib)^2 = (a^2 - b^2) + 2abi$. So $x = a^2 - b^2, y = 2ab$. End of proof of theorem. #

10.4 Exercise:

(1) Show that $(a^2 + b^2)(c^2 + d^2) = (ac - bd)^2 + (ad + bc)^2$

(2) There are three types of prime numbers: $2, p \in 4\mathbb{Z} + 1, q \in 4\mathbb{Z} - 1$. Show that a positive integer n can be expressed as certain $x^2 + y^2$ if and only if the indices of its prime factors of type $q \in 4\mathbb{Z} - 1$ are all even.

(3) Suppose $x^2 + y^2 = n$.
Explain why n determines $x + iy$ as follows:
  (i) If n contains the prime 2, with index M, then $x + iy$ contains
  $$(1 + i)^{M_1}(1 - i)^{M_2}, M_1 + M_2 = M$$
  There are $M + 1$ many possibilities.
  (ii) If n contains prime $p \in 4\mathbb{Z} + 1$, with index P, then $x + iy$ contains
  $$(a + ib)^{P_1}(a - ib)^{P_2}, P_1 + P_2 = P$$
  There are $P + 1$ many possibilities.
  Note that for any $p \in 4\mathbb{Z} + 1$, we pre-associate to it a fixed decomposition $p = (a + ib)(a - ib)$.
  (iii) If n contains prime $q \in 4\mathbb{Z} - 1$, with index $2Q$, then $x + iy$ contains
  $$q^Q$$
The whole $x + iy$ is the product of the above described factors, multiplied by a unit, which can be any of the four $\{\pm 1, \pm i\}$.

(4) Note that for any prime number p,

$$\frac{1}{1-1/p} = 1 + \frac{1}{p} + \frac{1}{p^2} + \frac{1}{p^3} + \cdots$$

$$\left(\frac{1}{1-1/p}\right)^2 = 1 + \frac{2}{p} + \frac{3}{p^2} + \frac{4}{p^3} + \cdots$$

(5) Explain why

$$4 \cdot \left(\frac{1}{1-1/2}\right)^2 \cdot \left(\prod_{p \in 4\mathbb{Z}+1} \frac{1}{1-1/p}\right)^2 \cdot \prod_{q \in 4\mathbb{Z}-1} \frac{1}{1-1/q^2} = \sum_{(x,y) \neq (0,0)} \frac{1}{x^2+y^2}$$

(6) Show that

$$\prod_{p \in 4\mathbb{Z}+1} \frac{1}{1-1/p} = +\infty$$

11 The integer equation $x^3 + y^3 = z^3$

The objective of this section is to show that integer equation
$$x^3 + y^3 = z^3$$
has no non-trivial solution.

Let $\omega = e^{2\pi i/3} = (-1 + \sqrt{3}i)/2$
Note that $x^3 - 1 = (x-1)(x-\omega)(x-\omega^2)$
Replacing x by $(-x/y)$:
$x^3 + y^3 = (x+y)(x+\omega y)(x+\omega^2 y)$
This is the reason why we want to concentrate on complex numbers $x + \omega y$, with x, y being integers.

11.1 $\mathbb{Z}[\omega]$ and its basic properties

$\mathbb{Z}[\omega]$ is defined to be the set of all the complex numbers $x + \omega y$, where x and y are integers.

For any $x + \omega y \in \mathbb{Z}[\omega]$, define its norm to be
$$N(x + \omega y) = (x + \omega y)(x + \bar{\omega} y) = x^2 - xy + y^2$$
This is a positive integer, unless $x + \omega y = 0$, i.e., $x = y = 0$, when the norm is zero. Again
$$N(\alpha\beta) = N(\alpha)N(\beta).$$

In $\mathbb{Z}[\omega]$, units are those elements α such that α^{-1} still in $\mathbb{Z}[\omega]$.
An element is a unit if and only if its norm is 1.

Proposition: In $\mathbb{Z}[\omega]$, there are six units:
$$\pm 1, \pm\omega, \pm\omega^2 = \mp(1+\omega).$$

Proof: Suppose $x + \omega y \in \mathbb{Z}[\omega]$ is unit.
Then $x^2 - xy + y^2 = 1$.

If x and y are of the same sign: $(x-y)^2 + xy = 1$
$x = 0, y = \pm 1$; $y = 0, x = \pm 1$; $x = y = \pm 1$.

If x and y are of different sign: $x^2 + (-xy) + y^2 = 1$
$x = 0, y = \pm 1$; $y = 0, x = \pm 1$. #

Like in $\mathbb{Z}[i]$, for a similar reason, there also exists an Euclid algorithm in $\mathbb{Z}[\omega]$. $\mathbb{Z}[\omega]$ would be (almost) useless without one.

Proposition: There exists an Euclid algorithm in $\mathbb{Z}[\omega]$ in the sense that, given any two elements $\alpha, \beta \in \mathbb{Z}[\omega]$, $\beta \neq 0$, the division α/β can be written as $\alpha/\beta = \gamma + \delta/\beta$, where $\gamma, \delta \in \mathbb{Z}[\omega]$, such that $N(\delta) < N(\beta)$.

The proof is similar to the one given for $\mathbb{Z}[i]$, but uses the following lemma instead, and is left to the reader.

Lemma: For any $z \in \mathbb{C}$, there exists $z_0 \in \mathbb{Z}[\omega]$, such that $|z - z_0| \leq \sqrt{3}/2$.

Proof: In the following figure, the graph on the left shows the distribution of $\mathbb{Z}[\omega]$ in the complex plane $\mathbb{C}$. The graph on the right then tells why the lemma is true, which implies the existence of an Euclidean algorithm because $\sqrt{3}/2 < 1$.

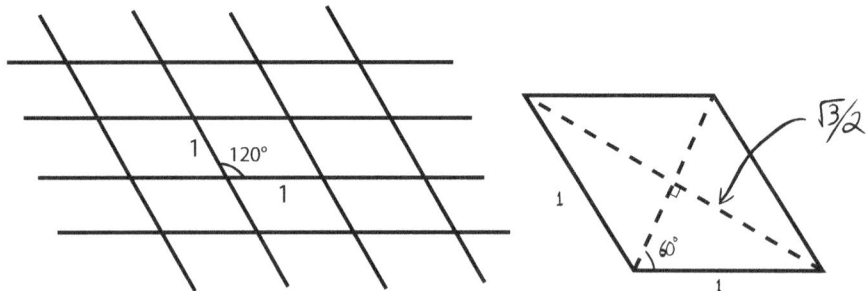

Primes in $\mathbb{Z}[\omega]$

In $\mathbb{Z}[\omega]$, an element is said to be prime if it cannot be further decomposed into non-trivial products of other elements of $\mathbb{Z}[\omega]$.

The existence of an Euclid algorithm ensures that primes in $\mathbb{Z}[\omega]$ behave just like the ordinary primes in integers. (See §10.1 for detailed discussions in case of $\mathbb{Z}[i]$.)

Similarity 1: If elements $\alpha, \beta \in \mathbb{Z}[\omega]$ are relatively prime, i.e., don't have non-trivial common factors, then there exist elements $\gamma, \delta \in \mathbb{Z}[\omega]$, such that $\gamma\alpha + \delta\beta = 1$.

Similarity 2: If α is a prime that divides $\beta\gamma$, then α must divide either β or γ.

Similarity 3: In $\mathbb{Z}[\omega]$, any element can be decomposed into the product of its prime factors, and the decomposition is essentially unique.

11.2 Identifying primes in $\mathbb{Z}[\omega]$

Note 1: If $\alpha \in \mathbb{Z}[\omega]$ such that $N(\alpha)$ is a prime number, then α is prime in $\mathbb{Z}[\omega]$: If α is not prime, then $\alpha = \beta\gamma$, where $\beta, \gamma \in \mathbb{Z}[\omega]$ are not units. Since $N(\beta)N(\gamma)$ is a prime number, one of $N(\beta), N(\gamma)$ must be 1, i.e., a unit. Contradiction.

Note 2: Given a prime number p, there are two possibilities: (1) p remains prime in $\mathbb{Z}[\omega]$. (2) p is no longer prime in $\mathbb{Z}[\omega]$, in which case, there exists a prime element $\alpha \in \mathbb{Z}[\omega]$ such that $N(\alpha) = p$, i.e., $p = \alpha\bar{\alpha}$. The reason is that, when $p \in \mathbb{Z}[\omega]$ no longer prime, it decomposes into a non-trivial product $p = \alpha\beta$. Then $N(\alpha)N(\beta) = p^2$. Since neither $N(\alpha)$ nor $N(\beta)$ can be 1, they must both be p.

$2 \in \mathbb{Z}[\omega]$ is prime:

Let's see if we can find $\alpha \in \mathbb{Z}[\omega]$ such that $N(\alpha) = 2$.
Let $\alpha = x + \omega y$.
Then $x^2 - xy + y^2 = 2$.
If x and y are of same sign:
$\quad (x-y)^2 + xy = 2$. So $x - y = \pm 1, xy = 1$. No solution.
If x and y are of different sign:

$x^2 + (-xy) + y^2 = 2$
One of these three terms must be zero.
So one of x, y is zero. No solution.
So no, we can't find any $\alpha \in \mathbb{Z}[\omega]$ such that $N(\alpha) = 2$.
So 2 remains prime in $\mathbb{Z}[\omega]$.

$3 \in \mathbb{Z}[\omega]$:

Let's see if we can find $\alpha \in \mathbb{Z}[\omega]$ such that $N(\alpha) = 3$.
Let $\alpha = x + \omega y$.
Then $x^2 - xy + y^2 = 3$.

If x and y are of same sign:
$(x-y)^2 + xy = 3$
Note that for $(x-y)^2$, 0 and 1 are the only possible values. So for xy, the possible values are 2 and 3. The solutions are $(x, y) = \pm(1, 2), \pm(2, 1)$.

If x and y are of different sign:
$x^2 + (-xy) + y^2 = 3$
If any of the three terms on the left of the equation is zero, then either x or y is zero, in which case the equation has no solution, because $\sqrt{3}$ is not an integer. So none of the three terms is zero. So all of them must be 1. The solutions are $x = \pm 1, y = \mp 1$.

So the solutions to $N(\alpha) = 3$ are
$\alpha = \pm(1 + 2\omega), \pm(2 + \omega), \pm(1 - \omega)$.
Actually, all of them are essentially the same:
$1 + 2\omega = \omega(1 - \omega) \sim 1 - \omega$
$2 + \omega = -\omega^2(1 - \omega) \sim 1 - \omega$

Conclusion:

$1 - \omega$ is a prime in $\mathbb{Z}[\omega]$
$3 = (-1/\omega)(1 - \omega)^2$

A general rule

Recall that $\mathbb{Z}_p$ denotes the set of all integers, but with a new understanding, that two integers n, m are considered the same if their difference $n - m$ is divisible by p. $\mathbb{Z}_p = \{0, 1, \cdots, p-1\}$, in which operations $+, -, \times, \div$ still work as usual, and is called a "field".

<u>Lemma:</u> Let p be a prime number. Then p remains prime in $\mathbb{Z}[\omega]$ if and only if the equation $x^3 = 1$ doesn't have any non-trivial solution in $\mathbb{Z}_p$.

Proof: Note that over $\mathbb{Z}_p$, $x^3 = 1$ doesn't have any non-trivial solution simply means $x^2 + x + 1$ is irreducible.

Suppose $x^2 + x + 1$ is reducible over $\mathbb{Z}_p$. Then
$x^2 + x + 1 = (x+a)(x+b)$, over $\mathbb{Z}_p$, i.e.,
$x^2 + x + 1 = (x+a)(x+b) + p \cdot \Box(x)$
Where a, b are integers, $\Box(x)$ is a polynomial with integer coefficients
Let $x = \omega$
$0 = (\omega + a)(\omega + b) + p \cdot \Box(\omega)$
So p divides $(\omega + a)(\omega + b)$. If p were prime in $\mathbb{Z}[\omega]$, then it would have to divide either $\omega + a$, or $\omega + b$. Not possible, because $(\omega+a)/p, (\omega+b)/p$ are not elements of $\mathbb{Z}[\omega]$. So p is not prime in $\mathbb{Z}[\omega]$.

Conversely, suppose p is not prime in $\mathbb{Z}[\omega]$. Then there exists $\alpha \in \mathbb{Z}[\omega]$ such that $N(\alpha) = p$. (α is thus a prime, $p = \alpha \bar{\alpha}$.) Write $\alpha = a + \omega b$, where a, b are integers. Then $a^2 - ab + b^2 = p$. Note that neither a nor b can have p as a prime factor.

Going down to $\mathbb{Z}_p$: $a^2 - ab + b^2 = 0$, neither a nor b is zero. $(a/b)^2 - a/b + 1 = 0$. So $x = -a/b$ is a root of $x^2 + x + 1 = 0$. So $x^2 + x + 1$ is reducible over $\mathbb{Z}_p$.   #

<u>Theorem:</u> Let p be a prime number. Then p remains prime in $\mathbb{Z}[\omega]$ if and only p is of type $3\mathbb{Z} - 1$.

Proof:

The case $p = 3$: Note that in $\mathbb{Z}_3$, all elements satisfy $x^3 = 1$. So by the preceding lemma, $3 \in \mathbb{Z}[\omega]$ is no longer prime.

The case $p \in 3\mathbb{Z}+1$: Write $p = 3n+1$. Then
$$x^{p-1} - 1 = x^{3n} - 1 = (x^3 - 1)(\cdots)$$
Recall that over $\mathbb{Z}_p$, the polynomial $x^{p-1} - 1$ completely splits, i.e., its set of roots are precisely the $(p-1)$ many elements of $\mathbb{Z}_p - 0$. See §9.3. We conclude that $x^3 - 1$ completely splits over $\mathbb{Z}_p$. So by the preceding lemma, $p \in \mathbb{Z}[\omega]$ is no longer prime.

The case $p \in 3\mathbb{Z}+2$:
Write $p = 3n+2$
In $\mathbb{Z}_p$,
$x^3 = 1 \Longrightarrow$
$x^{3n} = 1 \Longrightarrow$
$x^{p-2} = 1$
Note that $x^{p-1} \equiv 1$ for any $x \in \mathbb{Z}_p - 0$. So $x^{p-2} = 1$ would imply $x = 1$. This proves that the equation $x^3 = 1$ has no non-trivial solution over $\mathbb{Z}_p$. By the preceding lemma, p remains prime in $\mathbb{Z}[\omega]$. #

Examples:
These remain prime in $\mathbb{Z}[\omega]$: $2, 5, 11, 17, 23, \cdots$
These splits in $\mathbb{Z}[\omega]$: $3, 7, 13, 19, \cdots$

Remark: The prime numbers that don't split, together with the prime elements generated by the prime numbers that do split, constitute all the primes elements there are in $\mathbb{Z}[\omega]$. To see why, let $\mathfrak{p}$ be an arbitrary prime element in $\mathbb{Z}[\omega]$. Then $\mathfrak{p} \cdot \bar{\mathfrak{p}} = N(\mathfrak{p})$ is a positive integer, which decomposes into a product of prime numbers some of which further decompose in $\mathbb{Z}[\omega]$. Since $\mathfrak{p}$ is prime that divides the final version of the product, it must be equal to one of the prime elements in the final product, up to a possible multiplication by a unit.

Exercise:

(1) Explain how the following decompositions were obtained
$$\begin{aligned} 7 &= (1/\omega)(3+\omega)(1+3\omega) \\ &= (1/\omega)(3+2\omega)(2+3\omega) \\ &= (-1/\omega)(2-\omega)(1-2\omega) \end{aligned}$$

(2) Classify the six elements, listed above on the right, according to whether they are essentially the same. Note that in $\mathbb{Z}[\omega]$, two elements are essentially the same if they differ only by the multiplication of a unit.

Exercise: Suppose p is a prime number that splits in $\mathbb{Z}[\omega]$. Let $\alpha \in \mathbb{Z}[\omega]$ be a prime element such that $\alpha\bar{\alpha} = p$. Show that, with the exception of $p = 3$, α and $\bar{\alpha}$ are completely different, i.e., they are not essentially the same.

11.3 The integer equation $x^3 + y^3 = z^3$

We are now ready to deal with integer equation $x^3 + y^3 = z^3$, which decomposes into $(x+y)(x+\omega y)(x+\omega^2 y) = z^3$. Our strategy is to use prime elements of $\mathbb{Z}[\omega]$ to analyze the equation. It turns out that the most effective prime element here is $(1-\omega)$, the prime element that splits $3 \in \mathbb{Z}[\omega]$.

The first thing we do is to classify elements of $\mathbb{Z}[\omega]$ "up to $(1-\omega)$".

Lemma 1: Any element $\alpha \in \mathbb{Z}[\omega]$ can be expressed as $\alpha = \epsilon + (1-\omega)\mathcal{A}$, where $\epsilon \in \{0, \pm 1\}$, $\mathcal{A} \in \mathbb{Z}[\omega]$.

Proof: Write $\alpha = x + \omega y$, where x, y are integers. Then $\alpha = (x+y) - (1-\omega)y$. $x+y$ is an integer, so can be expressed as certain $\epsilon+3n$, where ϵ belongs to $\{0, \pm 1\}$, and n is integer. Note that $3 = (1-\omega)\cdot\square, \square \in \mathbb{Z}[\omega]$.
#

Lemma 2: Let $\alpha \in \mathbb{Z}[\omega]$. If α doesn't contain $(1-\omega)$ as a prime factor, then $\alpha^3 \in \{\pm 1\} + (1-\omega)^4\mathbb{Z}[\omega]$.

Proof:
$\alpha = \epsilon + (1-\omega)\mathcal{A}$, where $\epsilon \in \{\pm 1\}$, $\mathcal{A} \in \mathbb{Z}[\omega]$.
$\alpha^3 = \epsilon^3 + 3\epsilon^2(1-\omega)\mathcal{A} + 3\epsilon(1-\omega)^2\mathcal{A}^2 + (1-\omega)^3\mathcal{A}^3$

$\epsilon^3 = \epsilon \in \{\pm 1\}$
$3 = (-1/\omega)(1-\omega)^2$
Third term already in $(1-\omega)^4 \mathbb{Z}[\omega]$

So we concentrate on the sum of the second and forth terms
$3\epsilon^2(1-\omega)\mathcal{A} + (1-\omega)^3 \mathcal{A}^3$
$= (-1/\omega)(1-\omega)^2 \cdot 1 \cdot (1-\omega)\mathcal{A} + (1-\omega)^3 \mathcal{A}^3$
$= (1-\omega)^3(\mathcal{A}^3 - \mathcal{A}/\omega)$
$= (1-\omega)^3([\mathcal{A}/\omega]^3 - \mathcal{A}/\omega)$
$= (1-\omega)^3(\mathcal{A}/\omega)(\mathcal{A}/\omega - 1)(\mathcal{A}/\omega + 1)$

According to Lemma 1, the element $\mathcal{A}/\omega \in \mathbb{Z}[\omega]$ can be expressed as one of $\{0, \pm 1\}$ plus an element that contains $(1-\omega)$ as a factor. So one of the three terms in $(\mathcal{A}/\omega)(\mathcal{A}/\omega - 1)(\mathcal{A}/\omega + 1)$ must contain $(1-\omega)$. Thus, the above expression always contains $(1-\omega)^4$. #

<u>Theorem:</u> In $\mathbb{Z}[\omega]$, equation $\alpha^3 + \beta^3 = \gamma^3$ doesn't have any non-trivial solution.

Proof: Note that our so-called trivial solutions are the ones in which one of $\{\alpha, \beta, \gamma\}$ is zero. Assume that $\{\alpha, \beta, \gamma\}$ are all non-zero, such that $\alpha^3 + \beta^3 = \gamma^3$. We may further assume that $\{\alpha, \beta, \gamma\}$ are relatively prime. Obviously, $\{\alpha, \beta, \gamma\}$ are actually pairwise relatively prime.

We first point out that one of $\{\alpha, \beta, \gamma\}$ must contain $(1-\omega)$ as a factor: Assume that none of $\{\alpha, \beta, \gamma\}$ contains $(1-\omega)$. According to Lemma 2, $\{\alpha^3, \beta^3, \gamma^3\}$ all belong to $\{\pm 1\} + (1-\omega)^4 \mathbb{Z}[\omega]$.
So $\alpha^3 + \beta^3 - \gamma^3$ belongs to
$\{\pm 1\} + \{\pm 1\} - \{\pm 1\} + (1-\omega)^4 \mathbb{Z}[\omega]$
$\{\pm 1\} + \{\pm 1\} - \{\pm 1\} = \{\pm 1, \pm 3\}$
Note that 1 doesn't contain $(1-\omega)$, while 3 only contains $(1-\omega)^2$.
So $\alpha^3 + \beta^3 - \gamma^3$ can never be zero. Contradiction.

Since $\{\alpha, \beta, \gamma\}$ are pairwise relatively prime, only one of them contains $(1-\omega)$. We may well assume that it is the third. So we end up with equation
$$\alpha^3 + \beta^3 = U \cdot (1-\omega)^{3n} \cdot \gamma_0^3$$
Here, $\{\alpha, \beta, \gamma_0\}$ are pairwise relatively prime, none of which contains $(1-\omega)$ as factor. U is a unit. Integer $n \geq 1$.

We next point out that in fact $n \geq 2$: Since none of $\{\alpha, \beta, \gamma_0\}$ contains $(1-\omega)$ as factor, by Lemma 2, $\{\alpha^3, \beta^3, \gamma_0^3\}$ all belong to $\{\pm 1\} + (1-\omega)^4 \mathbb{Z}[\omega]$.

$$\alpha^3 + \beta^3 \in \epsilon + (1-\omega)^4 \mathbb{Z}[\omega], \epsilon \in \{\pm 1\} + \{\pm 1\} = \{0, \pm 2\}$$

$$U \cdot (1-\omega)^{3n} \cdot \gamma_0^3 \in U \cdot (1-\omega)^{3n} \cdot \left(\{\pm 1\} + (1-\omega)^4 \mathbb{Z}[\omega] \right)$$

Suppose $\epsilon \in \{\pm 2\}$: Note that $2 = -1 + 3 = -1 + (-1/\omega)(1-\omega)^2$. The above two terms can't be equal. So ϵ must be 0. Therefore $\alpha^3 + \beta^3$ contains $(1-\omega)^4$ at least. We see that, correspondingly, $3n \geq 4$. So $n \geq 2$.

$$(\alpha + \beta)(\alpha + \omega\beta)(\alpha + \omega^2\beta) = U \cdot (1-\omega)^{3n} \cdot \gamma_0^3$$

* Since $(1-\omega)$ is prime, one of $\{\alpha+\beta, \alpha+\omega\beta, \alpha+\omega^2\beta\}$ must contain $(1-\omega)$ as factor. Say $\alpha + \beta$ contains $(1-\omega)$. Then $\alpha + \omega\beta = \alpha+\beta-(1-\omega)\beta$ contains too. And $\alpha+\omega^2\beta = \alpha+\beta-(1-\omega)(1+\omega)\beta$ contains too. We conclude that all three of $\{\alpha+\beta, \alpha+\omega\beta, \alpha+\omega^2\beta\}$ contain $(1-\omega)$ as factor.

* Among $\{\alpha+\beta, \alpha+\omega\beta, \alpha+\omega^2\beta\}$, two of them contain only $(1-\omega)$, not $(1-\omega)^{\geq 2}$. Because if otherwise, then there are two, say $\{\alpha + \beta, \alpha + \omega\beta\}$, that contain $(1-\omega)^2$ as factor. Then $(1-\omega)^2$ divides the difference, i.e., $(1-\omega)\beta$. Then $(1-\omega)$ divides β. Contradiction.

* Any two of $\{\alpha + \beta, \alpha + \omega\beta, \alpha + \omega^2\beta\}$ have no common prime factors other than $(1-\omega)$: Because if $\mathfrak{p} \neq (1-\omega)$ is a prime that divides, say, $\alpha + \beta$ and $\alpha + \omega\beta$, then it divides the difference, i.e., $\mathfrak{p}$ divides $(1-\omega)\beta$. So $\mathfrak{p}$ divides β. Since $\mathfrak{p}$ divides $\alpha+\beta$, it divides α too. Contradiction, because α and β are relatively prime.

* To sum up: In $\{\alpha + \beta, \alpha + \omega\beta, \alpha + \omega^2\beta\}$, two contain $(1-\omega)$ exactly, the third contains $(1-\omega)^{3n-2}$ exactly. And if we get rid of the $(1-\omega)$ factor from all of them, then the three becomes pairwise relatively prime.

    For convenience, assume $\alpha + \omega^2\beta$ contains $(1-\omega)^{3n-2}$ as factor.

$$\frac{\alpha+\beta}{1-\omega} \cdot \frac{\alpha+\omega\beta}{1-\omega} \cdot \frac{\alpha+\omega^2\beta}{(1-\omega)^{3n-2}} = U \cdot \gamma_0^3$$

The three terms on the left are pairwise relatively prime. On the right is essentially, i.e., up to a multiplication by a unit, a third power to an element γ_0 in $\mathbb{Z}[\omega]$. By going through the primes in γ_0, one quickly conclude that all the three terms on the left must also be, essentially, the third power of elements in $\mathbb{Z}[\omega]$.

> An integer example to explain why:
>
> Suppose $abc = d^3$, where a, b, c, d are all positive integers. $\{a, b, c\}$ are pairwise relative prime, i.e., each of the pairs $\{a, b\}$, $\{b, c\}$, $\{c, a\}$ are relatively prime. Then a, b, c must all be the third powers of some other positive integers:
>
> For example, say $d = 5^6 \cdot 7^8 \cdot 11^{12}$. So $(5^6)^3$ divides abc. Since $\{a, b, c\}$ are pairwise relative prime, only one of them contains $(5^6)^3$. Similarly, $(7^8)^3$ is contained in just one of $\{a, b, c\}$. Same for $(11^{12})^3$. We conclude that there exist positive integers $\{a_0, b_0, c_0\}$ such that $a = a_0^3, b = b_0^3, c = c_0^3$.

Write
$$\begin{cases} \alpha+\beta = A \cdot (1-\omega) \cdot \mu^3 \\ \alpha+\omega\beta = B \cdot (1-\omega) \cdot \nu^3 \\ \alpha+\omega^2\beta = C \cdot (1-\omega)^{3n-2} \cdot \xi^3 \end{cases}$$

A, B, C are units. μ, ν, ξ are pairwise relatively prime, none containing $(1-\omega)$. Solve the first two equations for α and β, then enter them into the third:

$\alpha = B\nu^3 - \omega A\mu^3$
$\beta = A\mu^3 - B\nu^3$
$\alpha + \omega^2\beta = A\omega(\omega-1)\mu^3 + B(1-\omega^2)\nu^3$
$A\omega(\omega-1)\mu^3 + B(1-\omega^2)\nu^3 = C(1-\omega)^{3n-2}\xi^3$

$\mu^3 + X\nu^3 = Y(1-\omega)^{3(n-1)}\xi^3$
X, Y are units.

Let's show that X can only be ± 1: Since none of $\{\mu, \nu, \xi\}$ contains $(1-\omega)$, by Lemma 2, their third powers all belong to $\{\pm 1\} + (1-w)^4\mathbb{Z}[\omega]$.

The term $\mu^3 + X\nu^3$ belongs to
$\{\pm 1\} + X\{\pm 1\} + (1-\omega)^4 \mathbb{Z}[\omega]$
$\{\pm 1\} + X\{\pm 1\}$ has the following possibilities
$$\pm(1+1) = \pm 2 = \pm(3-1) = \mp 1 \pm (-1/\omega)(1-\omega)^2$$
$$\pm(1-1) = 0$$
$$\pm(1+\omega) = \mp \omega^2$$
$$\pm(1-\omega)$$
$$\pm(1+\omega^2) = \mp \omega$$
$$\pm(1-\omega^2) = \pm(1+\omega)(1-\omega) = \mp \omega^2 (1-\omega)$$
On the other hand, the term $Y(1-\omega)^{3(n-1)}\xi^3$ contains $(1-\omega)^3$ at least, because $n \geq 2$. We see that, for these two terms to be equal, the second possibility above must be the one that actually happens. Anyway, X must be ± 1. Note that if $X = -1$, we can make it 1 by simply replacing ν by $-\nu$.

$\mu^3 + \nu^3 = Y(1-\omega)^{3(n-1)}\xi^3$
$\{\mu, \nu, \xi\}$ are pairwise relatively prime, none containing $(1-\omega)$. Y is a unit. But n is reduced to $n-1$.

Eventually, n will be reduced to an unacceptable value.
Contradiction.
This proves the theorem. #

<u>Corollary</u>: The integer equation $x^3 + y^3 = z^3$ has no non-trivial solution.

12 The Fibonacci numbers

$$\cdots\ -8\quad 5\quad -3\quad 2\quad -1\quad 1\quad \underset{\underset{F_0}{\uparrow}}{0}\quad \underset{\underset{F_1}{\uparrow}}{1}\quad \underset{\underset{F_2}{\uparrow}}{1}\quad \underset{\underset{F_3}{\uparrow}}{2}\quad \underset{\underset{F_4}{\uparrow}}{3}\quad 5\quad 8\quad 13\ \cdots$$

These numbers satisfy
$F_n + F_{n+1} = F_{n+2}$

Exercise:
Let $y = \sum_{n=0}^{\infty} F_n x^n$

Show that $y = \dfrac{x}{1 - x - x^2}$

Deduce that
$$F_n = \frac{1}{\sqrt{5}}\left[\left(\frac{1+\sqrt{5}}{2}\right)^n - \left(\frac{1-\sqrt{5}}{2}\right)^n\right]$$

13 Continued fractions

13.1

A continued fraction is an expression

$$a + \cfrac{1}{b + \cfrac{1}{c + \cfrac{1}{\ddots}}}$$

The first value is an integer.
The rest of the values are positive integers.
It is also denoted as $[a; b, c, \cdots]$

Example: Find the continued fraction expression of 5/3.

$$5/3 = 1 + \frac{2}{3} = 1 + \frac{1}{3/2} = 1 + \cfrac{1}{1 + \cfrac{1}{2}}$$

So $5/3 = [1; 1, 2]$

Exercise: Find the continued fraction expression of 225/157

Example: Find the continued fraction expression of $\sqrt{2}$.

$$\sqrt{2} = 1 + (\sqrt{2} - 1) = 1 + \frac{1}{\sqrt{2}+1} = 1 + \frac{1}{2 + (\sqrt{2}-1)} = 1 + \frac{1}{2 + \frac{1}{\sqrt{2}+1}} = \cdots$$

$$= 1 + \cfrac{1}{2 + \cfrac{1}{2 + \cfrac{1}{2 + \cfrac{1}{\ddots}}}}$$

$$= [1; 2, 2, 2, \cdots]$$

Exercise: Verify that
$\sqrt{3} = [1; 1, 2, 1, 2, 1, 2, \cdots]$

Simple Facts

(1) Any number has a unique continued fraction expression.
(2) A number is rational if and only if its continued fraction expression is finite: If the expression is finite, then the number is a fraction and so is rational. On the other hand, a rational number is a fraction. When generating the continued fraction expression, each iteration leads to a strictly smaller denominator. All the denominators are positive integers. So only a finite number of iterations are possible.

Example:

$$\pi = 3 + \cfrac{1}{7 + \cfrac{1}{15 + \cfrac{1}{1 + \cfrac{1}{292 + \cfrac{1}{\ddots}}}}} = [3; 7, 15, 1, 292, \cdots]$$

$$\pi \approx 3 + \cfrac{1}{7 + \cfrac{1}{16}} = \frac{355}{113} = 3.1415929\cdots$$

$$\pi \approx 3 + \cfrac{1}{7 + \cfrac{1}{15 + \cfrac{1}{1 + \cfrac{1}{292}}}} = \frac{103993}{33102} = 3.1415926530\cdots$$

($\pi = 3.1415926535\cdots \leftarrow$ the exact value)

Example: $1 + \cfrac{1}{1 + \cfrac{1}{1 + \cfrac{1}{1 + \cfrac{1}{\ddots}}}} = ?$

Solution:
Let it be x

$x = 1 + \cfrac{1}{1 + x}$

$x = \cfrac{1 \pm \sqrt{5}}{2} = \cfrac{1 + \sqrt{5}}{2}$

13.2

Consider an infinite continued fraction

$$I_1 + \cfrac{1}{I_2 + \cfrac{1}{I_3 + \cfrac{1}{\ddots}}}$$

We want to answer two questions:
* Is it really convergent?
* If yes, how fast?

We sum up the answers in the following theorem. Details, not too complicated, are left to the reader as a nice exercise.

<u>Theorem:</u> Let $[I_1 I_2 I_3 \cdots]$ be an infinite continued fraction, where I_1 is an integer, $I_2, I_3, \cdots$ are positive integers.

(1) Let $x_n = [I_1 \cdots I_n]$, $n \geq 1$.
These values are distributed as follows

$$x_1 < x_3 < x_5 < \cdots\cdots < x_6 < x_4 < x_2$$

(2) The difference formula

$$x_2 - x_1 = \frac{1}{I_2}$$

$$x_3 - x_2 = -\frac{1}{[I_2 I_3][I_3]} \cdot \frac{1}{[I_2]}$$

$\cdots$

$$x_{n+1} - x_n = (-1)^{n-1} \cdot \frac{1}{[I_2 \cdots I_{n+1}][I_3 \cdots I_{n+1}] \cdots [I_{n+1}]} \cdot \frac{1}{[I_2 \cdots I_n][I_3 \cdots I_n] \cdots [I_n]}$$

(3) On the product $[I_2 \cdots I_n][I_3 \cdots I_n] \cdots [I_n]$, $n \geq 2$:
* The product is a sum of positive integers
* The number of terms in the sum is the Fibonacci number F_n
* The product $\geq F_n$

"$\geq$" becomes "=" only when $I_2 = I_3 = \cdots = 1$

(4) $|x_{n+1} - x_n| \leq \dfrac{1}{F_n F_{n+1}}$, $n \geq 1$.

In particular, the sequence $\{x_n\}$ is indeed convergent.

Remark:

$$\frac{1+\sqrt{5}}{2} = 1 + \cfrac{1}{1 + \cfrac{1}{1 + \cfrac{1}{1 + \cfrac{1}{\ddots}}}}$$

Among all continued fractions, this is the one where the speed of convergence is slowest.

Table of Contents of Part II

1 Two-dimensional surfaces
- 1.1   The 2-dimensional sphere S^2
- 1.2   The Möbius band
- 1.3   The projective space P^2
- 1.4a The concept of transversal intersection
- 1.4b One-sided surfaces in the 3-dimensional space?
- 1.5   The projective space in the 4-dimensional space
- 1.6   Why the name "projective space"
- 1.7   The torus T^2
- 1.8   The Klein bottle
- 1.9   The handlebody structure of a surface
- 1.10 The classification of surfaces

2 Cell complexes
3 The Euler number
4 The Gauss-Bonnet formula
5 Spherical geometry

Appendix

6 Appendix: Hyperbolic geometry
- 6.1   The unit hyperboloid
- 6.2   The hyperbolic functions
- 6.3   The hyperbolic inner product
- 6.4   The hyperbolic lines
- 6.5a Orthogonal and hyperbolic transformations
- 6.5b 2-dimensional orthogonal transformations
- 6.5c 2-dimensional hyperbolic transformations
- 6.5d 3-dimensional hyperbolic transformations
- 6.6   Hyperbolic transformations on the unit hyperboloid
- 6.7a Projecting the unit hyperboloid to the xy-plane

- 6.7b The hyperbolic plane $\mathbb{H}^2$
- 6.8   Hyperbolic transformations for $\mathbb{H}^2$
- 6.9   The hyperbolic distance
- 6.10 The hyperbolic angle
- 6.11 Formulas in hyperbolic geometry

7 Appendix: Hyperbolic geometry on surfaces

Index

Convex function, 2
Cubic equation, 17
    discriminant, 21
    elementary solution, 31, 33

Elementary construction, 31, 33
    $\cos(2\pi/17)$, 48
    $\cos(\pi/7)$, 42
    $\sin 10°$, 33
Euclid algorithm
    polynomial, 26
    $\mathbb{Z}[\omega]$, 64
    $\mathbb{Z}[i]$, 56

Fermat's Little Theorem, 51
Fibonacci numbers, 75, 80
Field, 25
    extension, 27, 28
    $\mathbb{Z}_p$, 50

Irreducible polynomial, 26

$n!$, 3, 11
Norm
    $\mathbb{Z}[i]$, 55
    $\mathbb{Z}[\omega]$, 63

Primes
    $\mathbb{Z}[\omega]$, 64, 65
    $\mathbb{Z}[i]$, 56, 59
Primes,infinite many
    $\mathbb{Z}$, 49

$4\mathbb{Z} + 1$, 53
$4\mathbb{Z} - 1$, 52

$\sin 10°$, 19, 20
    elementary construction, 33

Units
    $\mathbb{Z}[\omega]$, 63
    $\mathbb{Z}[i]$, 55

www.ingramcontent.com/pod-product-compliance
Lightning Source LLC
Chambersburg PA
CBHW081608220526
45468CB00010B/2815